CLIMATE CHAOS

CLIMATE CHAOS

LESSONS ON SURVIVAL
FROM OUR ANCESTORS

BRIAN FAGAN *and*
NADIA DURRANI

PUBLICAFFAIRS
New York

PublicAffairs
Hachette Book Group
1290 Avenue of the Americas, New York, NY 10104
www.publicaffairsbooks.com
@Public_Affairs

Printed in the United States of America

First Edition: September 2021

Published by PublicAffairs, an imprint of Perseus Books, LLC, a subsidiary of Hachette Book Group, Inc. The PublicAffairs name and logo is a trademark of the Hachette Book Group.

The Hachette Speakers Bureau provides a wide range of authors for speaking events. To find out more, go to www.hachettespeakersbureau.com or call (866) 376-6591.

The publisher is not responsible for websites (or their content) that are not owned by the publisher.

Print book interior design by Jeff Williams.

Library of Congress Cataloging-in-Publication Data

Names: Fagan, Brian M., author. | Durrani, Nadia, author.
Title: Climate chaos: lessons on survival from our ancestors / Brian Fagan and Nadia Durrani.
Description: First edition. | New York: PublicAffairs, 2021. | Includes bibliographical references and index.
Identifiers: LCCN 2021016383 | ISBN 9781541750876 (hardcover) | ISBN 9781541750883 (ebook)
Subjects: LCSH: Paleoclimatory—Research. | Climatic changes—History.
Classification: LCC QC884.2.C5 F34 2021 | DDC 304.2/5—dc23

LC record available at https://lccn.loc.gov/2021016383

ISBNs: 9781541750876 (hardcover); 9781541750883 (ebook)

LSC-C

Printing 1, 2021

To Michael McCormick
With thanks for encouragement and sage advice.
He is a paragon of the "dawning age
of consilient interdisciplinary inquiry of the
human and natural past."

CONTENTS

PREFACE

Nekhen, Upper Egypt, c. 2180 BCE. Ankhtifi was a very powerful man in an Egypt plagued by dissent and famine. He was a nomarch, a provincial governor, at least theoretically subordinate to the pharaoh but in fact one of the most influential men in the state. He walked in formal procession to the sun god Amun's temple, surrounded by well-armed guards. He wore white, his wig in perfect order, necklaces of semiprecious stones at his neck. The great lord looked neither left nor right in the bright sunlight, seemingly ignoring the silent, hungry crowds gathered along the route. He carried his long staff of office and a ceremonial mace, a richly decorated, knotted belt around his waist. The soldiers' eyes flickered back and forth, alert for spears and knives. The people had empty stomachs; the rations they received were meager; stealing and petty violence were on the rise. A horn sounded as the great man entered the temple with its dark shrine where the sun god awaited. Silence fell as the nomarch made offerings to Amun and prayed for a plentiful flood, for relief from recent years.

It had been thus for several generations, longer than many local farmers could recall. Down by the Nile, the priests had been watching the flood for days, marking its rise against steps in the bank. Some of them shook their heads, for they sensed that the flood was slowing. But they were hopeful, for they believed that the gods controlled the river and the floods that nourished it far upstream. Ankhtifi was a strong, forthright leader who ruled his people with an iron hand. He rationed food, controlled peoples' movements, and closed his province's borders, but this able and forceful man knew in his heart of hearts that he and his people were at the mercy of the gods. It had always been so.

Ankhtifi and his contemporaries lived in an Egyptian world encompassed by the Nile. He lived in troubled times, beset by poor river floods and hunger, which threatened the state's existence, not so different from our world today. The climatic stakes of our time are global and of unprecedented severity. As numerous people, from politicians and religious leaders to grassroot activists and scientists, have proclaimed, humanity's future is at stake. Many experts remind us that we have one more chance to correct humankind's course, to avoid potential extinction. This is true, but we've largely forgotten the vast legacy surrounding humans and climate change that we've inherited.

There's a common assumption that the human experience with ancient climatic shifts is irrelevant to today's industrialized world. Nothing could be further from the truth. We don't necessarily learn from the past directly. But through years of archeological study, we've learned more about ourselves, as individuals and as a society. And we've also come to understand a great deal more about the challenges of adapting to climate change over long periods of time.

Unfortunately, our promiscuous dependence on carbon-generating fossil fuels continues virtually unabated. The catastrophic wildfires that ravaged the American West in 2020 offer shattering proof of the menace posed by human-caused climate change. Persistent warming, a higher frequency of hurricanes and other extreme weather events, rising sea levels, unprecedented droughts, and record-breaking temperatures: the list of threats seems unending. A tidal wave of basic scientific research has established beyond all doubt that we humans are the cause of high carbon levels in the atmosphere and global warming.

Despite this research, platoons of climate change deniers, often funded by the very industries they are defending, proclaim that global warming, sea level rises, and the increasingly frequent extreme climatic events of today are part of the natural cycle of things. The "skeptics" spend vast sums of money on elaborate disinformation campaigns, even on conspiracy theories denigrating science. They are so convincing that a significant percentage of US citizens think they speak the truth. But on what basis do they draw such conclusions? Our concern here is with the dramatic advances in our knowledge of how humans have dealt with climate change over the past 30,000 years. How did people cope with such uncertainties of weather and climate? Which of their measures worked and which failed? What lessons can we take from their lives to inform ourselves and our future decisions? The claims of climate change deniers have no place in these discussions.

Even a quarter century ago, it would have been impossible to tell this story. Among all the historical sciences, archeology is unique in its ability to study human societies as they develop and change over enormously long periods. Archeologists' historical perspective extends back much further than the Declaration of Independence or the Roman Empire. The 5,100 years or so of written history are but a blink of an eye in terms of the 6 million years of human experience. In these pages, we focus our historical telescope on people and climate change in a segment of this long chronicle, the past 30,000 years from the height of the last Ice Age to modern times, a period of remarkable transformation in human society. A major revolution in paleoclimatology, the study of ancient climate, has subsequently changed our knowledge of ancient climates. Much of this research is highly specialized, technical, and very fast moving, with important research papers appearing weekly. To master it is a daunting task, which appeals to few laypeople. Rather than submerge our narrative in a morass of scientific detail, we've written what we call a prolegomenon on the climatology, which appears on p. 1. This attempts to provide an overview of major climatic phenomena (for example, El Niños and the North Atlantic Oscillation) and the most widely used methods for studying ancient climate either directly or using what are called proxies—more indirect approaches. We felt it best to provide a separate discussion of these subjects, lest we divert from the major thrust of our narrative, which is predominantly archeological and historical.

For the first time, we archeologists and historians can really begin to tell the story of ancient climate change. We believe that how humans of the past adapted to the repercussions of long- and short-term climate change has direct relevance to today's concerns with human-caused (anthropogenic) global warming. Why? Because we can learn from the lessons of the past: how our ancestors dealt with the difficulties arising from climate change—or did not. As astrophysicist Carl Sagan said in 1980, "You have to know the past to understand the future."

Climate Chaos draws not only from the latest paleoclimatology but also from new and often highly innovative research across a broad range of the humanities and human sciences, among them anthropology, archeology, ecology, and environmental history. We also offer you the contributions, often hidden away in specialized periodicals and university libraries, of those who have conducted in-depth inquiry into human behavior in relation to ancient climate over the past two decades. We have trawled these archives to bring to life the human response to climatic events of the past.

A 30,000-YEAR NARRATIVE

This is not a scientific textbook about ancient climate change; it's a story of how our ancestors adapted to its myriad shifts, large and small. The science of climate change is the background for the unfolding human narrative that we tell in these pages about the people of the past, the individuals who made up widely diverse societies—whether hunters and foragers, farmers and herders, or people living in preindustrial civilizations. These stories span thousands of years of human experience in times before government offices, weather forecasts, global models, satellites, or any of the technology that we take for granted today (see chronological table on p. xxi).

We begin the story during the late Ice Age, about 30,000 years ago. And so we should, because adaptations to extreme cold in clothing, technology, and strategies of risk management continued in use for thousands of years thereafter. Ice Age art, especially on cave walls, provides a powerful history of the complex human relationship with the natural world that survived, albeit in different guises, into the modern world. The last glacial maximum reached its peak about 18,000 years ago, followed by prolonged, irregular natural global warming. The adaptive skills of late Ice Age people were a vibrant legacy for the rapidly changing, warming world that confronted their successors after 15,000 years ago. We soon discover one of the realities of climate change: its volatility. It surrounds humanity on every side, ebbing and flowing in cycles of cold and greater warmth, rainfall and flood, harsh droughts long and short, and climatic shifts sometimes triggered by major volcanic eruptions.

The first three chapters cover a slice of time between the end of the Ice Age some 15,000 years ago and the first millennium CE. Momentous times these were—the shift from hunting and foraging to agriculture and animal husbandry took place, followed shortly by the emergence of the first preindustrial urban civilizations. Intuition and social memory were crucial for the success of subsistence agriculture, where experience was always a vital part of risk management and adaptation, coupled with an intimate knowledge of the local environment. However, increasingly complex and socially stratified societies soon acquired not only dramatic social inequality but a growing vulnerability to rapid climate change. By moving large populations into cities and making their inhabitants dependent on food rations from the state, rulers in turn depended heavily on grain surpluses from city hinterlands and on intensified agriculture controlled by the political elite. The risks swelled with the growth of cities like Rome and Constantinople, which came to depend heavily on grain imported from distant lands like Egypt and North Africa. They also became increasingly vulnerable to plague pandemics like the catastrophic Justinian Plague of 541 CE.

Chapters 4 to 10 carry us later into the first millennium CE, to the end of the Roman Empire, the rise of Islam in the Middle East, and the height of the Maya world in Central America. Here, the climatological record becomes much more fine-grained. We return to the increasing, sometimes fatal, vulnerability of complex, centralized preindustrial states. The great city of Angkor Wat in Cambodia dissolved when its elaborate waterworks came under stress. Cores extracted from ice sheets and lakes in the southern Andes chronicle the rise and collapse—the word is appropriate here—of the Tiwanaku and Wari states in the Bolivian and Peruvian highlands over 1,000 years ago. Strong and weak monsoons either supported or undermined civilizations in Southeast and South Asia and affected volatile kingdoms in Southern Africa.

The stories of the diverse preindustrial civilizations described in these seven chapters make an important point about ancient climate change generally. Long- or short-term climatic shifts never "caused" an ancient civilization to collapse. Rather, they were a major player in fostering dangerous levels of ecological, economic, political, and social vulnerability in societies with authoritarian leadership blinkered by rigid ideologies. Think of a pebble cast into a calm pond and the ripples that radiate outward from the point of impact. Those ripples are the economic and other factors that come together to tear apart the seeming calm of prosperous states.

Thereafter, we experience the more familiar climatological and historical territory of the past 1,300 years, the climatic yo-yo that included the Medieval Climate Anomaly and the Little Ice Age, both of which we will discuss in Chapters 11 to 14. Again, our perspective is global, concerned with how climate change affected major events, like Europe's Great Famine of 1315 to 1321 and the Black Death of 1351, as well as the impact of reduced sunspot activity, including the well-known Maunder Minimum of 1545 to 1715. We describe the impact of cold on the Jamestown colonists in North America, how Ancestral Pueblo Indians in the American Southwest adapted to prolonged periods of megadrought, and how climate facilitated the so-called Golden Age of the Netherlands, whose shrewd merchants and mariners used prevailing easterly winds from cold conditions for ocean voyaging. Chapter 14 describes the celebrated "Year Without a Summer" of 1816, caused by the Mount Tambora eruption of the preceding year, which had global impacts and caused major famines. And finally, we reach the global warming that began in the late nineteenth century, caused by rising industrial pollution.

An interesting historical journey, but what does all this mean to us? Chapter 15 stresses that humanity's past experience with long- and short-term climate change is vital to living with today's unprecedented anthropogenic warming. Here, we carefully lay out the differences between the past and present, notably in the scale of the climatic problem. Predictions of climatic Armageddon abound in books of all kinds, to the point that they often sound like modern-day versions of the biblical book of Revelation, complete with the Four Horsemen of the Apocalypse. In contrast, we argue that there are important lessons to learn from traditional societies, ancient and still thriving. For example, our approach to climate change must involve long-term planning and fiscal management, something that was unknown in ancient times, except in Andean societies, which knew the realities of living with long-term drought. We've learned that a great deal depends, even today, on *local* responses to threatening climate change, as well as on international cooperation on scales unimaginable in the past.

PRESENTS FROM THE PAST

The people of the past have bequeathed us priceless lessons about adapting to climate change. But first a fundamental point: we are human beings, just like our forebears, and have inherited the same brilliant qualities of forward thinking, planning, innovation, and cooperation. We are *Homo sapiens*, and these qualities have always helped us adapt to climate changes. They are a priceless legacy of experience.

A second gift from the past: an abiding reminder that kin ties and the innate human capacity for cooperation are valuable assets, even in densely populated megacities. One must only look at ancient or modern Pueblo society in the American Southwest to realize that kinship, obligations to one another, and mechanisms that break down isolation remain an essential glue for human societies under stress. Today, we can see those same relationships in various community groups, be they churches or clubs. Kin ties are one coping mechanism. So are the strategies of dispersal and mobility, which for thousands of years were highly adaptive ways of coping with drought or destruction wrought by sudden flooding. Mobility in the form of involuntary migration is still a significant human reaction to climate change: witness the thousands of people fleeing drought in Northeast Africa or trying to move northward into the United States. Today, we talk about ecological refugees. We are actually witnessing the ancient survival strategy of mobility on a truly massive scale.

There's far more. The societies of the past lived in intimate association with their environments. They never had the benefit of scientific weather forecasting, let alone computer models or even one of the many proxies now at our disposal. The Babylonians and others, including medieval astronomers, consulted the heavenly bodies without success. Until the nineteenth century, even the most expert climatic predictions involved local phenomena such as cloud formations or sudden temperature changes. Farmers and city dwellers alike relied on subtle environmental signposts acquired over many generations, such as the dense cloud formations that foreshadow approaching hurricanes. Similarly, fisherfolk and mariners identified the subtle changes in ocean swells that arrived ahead of powerful storms. The experience of the past reminds us that adaptations to climate change are, more often than not, local initiatives, based on local experience and understanding. Such adaptations, be they the building of a seawall, moving houses to higher ground, or communal response in the face of a catastrophic flood, rely on local experience and environmental knowledge. Most ancient societies, whether small villages or cities, were well aware that they were at the mercy of climatic forces, not in control of them.

Looking back over the millennia, we can distinguish general categories of climate change that challenged our forebears. Catastrophic events, such as exceptionally strong El Niños along the Peruvian coast or massive volcanic eruptions with their destructive ash clouds, which ruined crops, were short-lived but caused suffering and sometimes serious damage and loss of life. Yet once the event ended, normal climatic conditions returned, and the victims recovered. The effects were usually short-term and soon done with, often within the span of a single lifetime. Recovery from such climatic punches required cooperation, close ties, and strong leadership, an enduring legacy from the past.

In small-scale societies, leadership fell to kin leaders and elders, to individuals with experience and the kind of personal charisma that engendered loyalty. Much depended on reciprocal obligations between fellow kin and also on leaders' ability to control and manipulate food surpluses.

Climatic events are different propositions from short-term climatic shifts: a prolonged drought cycle, a rainy decade, or persistent flooding that decimates growing crops. Many subsistence farming societies of the past, like the Moche and Chimu of coastal Peru, were only too aware of the hazards of prolonged droughts. They relied on mountain runoff from the Andes to nourish carefully designed irrigation works in desert river valleys headed toward the Pacific. Both also depended on the rich coastal anchovy fisheries for much of their diet and on groundwater and carefully maintained irrigation canals to distribute water supplies in an environment where every drop counted. Their resilience depended on community-based stewardship of water systems, supervised by powerful chiefs.

It was no coincidence that the preindustrial civilizations of the past 5,000 years thrived on social inequality, as societies run for the benefit of the few. Everything depended on carefully acquired and maintained food surpluses, for societies like ancient Egypt and the Khmer civilization in Southeast Asia fed both nobles and commoners with rations. Village farmers living and working close to the land could survive short droughts by relying on less popular crops or perhaps wild plant foods. There might be hunger, but life would go on. Prolonged dry cycles, like the celebrated megadrought of 2200 to 1900 BCE, often called the 4.2 ka Event, which spread across the eastern Mediterranean and South Asia, were another matter. Confronted with this drought, the pharaohs were unable to feed their people. The state dissolved into competing provinces. The most successful of provincial leaders, familiar with solving problems on a local scale, managed to feed their people and restrict movement. There was no more talk of divine pharaohs controlling Nile floods. Later kings invested heavily in irrigation, and ancient Egypt survived until Roman times.

Nor was it a coincidence that preindustrial civilizations were, for the most part, volatile entities that rose and fell with bewildering rapidity. Much depended on the ability of rulers to move grain and essential commodities over considerable distances. The pharaohs had the Nile at their doorsteps, whereas the Maya and many Chinese and Mesopotamian states had to rely on human labor and pack animals. In political terms this meant that adapting to climate change was once again a local matter, since infrastructure limitations were such that most rulers could only tightly control territories about one hundred kilometers across. Waterborne bulk transportation was the answer. The Roman emperors fed thousands of their subjects on Egyptian and North African grain, but their climatic vulnerability to crop failure in remote lands increased a hundredfold.

With the growth of industrialization, the development of steam power, and the accelerating globalization of the nineteenth to the twenty-first centuries, the complexities of larger-scale societies have made adapting to climate change a far more challenging undertaking. But there is hope for the future, and part of this optimism comes from our brilliant ability as humans to seize opportunities and adapt to climate change on a massive scale. And the lessons from the past provide us with encouraging perspectives on the future.

Decisive leadership and that most central human quality, our ability to cooperate with one another, are two fundamental, long-established strategies the past brings to the climatic table. Human nature and our responses to change and sudden emergencies are sometimes very predictable. The Pompeii eruption and other disasters chronicle relatable behavior in the face of catastrophic events. We are all the same species, and we have much to learn from one another—and from our collective past. If not now, then soon, humanity will have to shift difficult gears, for the ultimate reality is that one day, perhaps tomorrow, perhaps in several centuries, humankind will confront a climatic disaster that transcends petty nationalisms and affects all of us—simultaneously, and as profoundly as a pandemic. With this book, we intend to unpack the past to help readers navigate the present and to use ancient insights to move onward into the future.

AUTHORS' NOTE

DATES

All radiocarbon dates are calibrated against calendar years. The BCE/CE convention is used throughout. Dates earlier than 10,000 BCE are quoted as "years ago."

PLACE-NAMES

Modern place-names reflect the current most commonly used spellings. Where appropriate, widely accepted ancient usages are employed.

MEASUREMENTS

All measurements are given in metric, this now being common scientific convention.

MAPS

In some instances, locations that are obscure or minor or that are in or very close to modern cities have been omitted from the maps.

CHRONOLOGICAL TABLE

A generalized chronological table follows. Given the broad timescale of the book and sometimes inevitable abrupt shifts between centuries and millennia, chronological information is also given with each chapter title and accompanies many of the subheads.

MAJOR CLIMATIC AND HISTORICAL EVENTS, 15,000 YEARS AGO TO PRESENT

This table lists some of the major climatic and cultural developments since the Ice Age. It makes no attempt to be comprehensive. Major climatic events are in **bold**. Dates before 10,000 BCE are listed in years before the present.

CE

2020	**Continued anthropogenic warming.**
1875–1877	Severe droughts in India and northern China. Millions perish.
c. 1850	**Anthropogenic warming takes serious hold as black carbon emissions skyrocket. The Little Ice Age peters out.**
1817–1830s	Major cholera epidemic kills thousands.
1816	The "Year Without a Summer" brings volcano-caused cold.
	Mary Shelley writes *Frankenstein.*
1815	Mount Tambora erupts.
c. 1760	So-called Industrial Revolution begins.
1645–1715	**Maunder Minimum.**

1607	Jamestown founded. Drought from 1606 to 1612 causes suffering.
1602	Dutch East India Company founded.
1600	Mount Huaynaputina, Peru, erupts.
1600s	Nunalleq, Alaska, occupied.
1565	St. Augustine, Florida, founded.
1584–1586	Roanoke Island settled and abandoned.
1590	King Henri IV of France besieges Catholic Paris.
1560s–1620	**Grindelwald Fluctuation.**
1458	Mount Kuwae, Vanuatu, southwestern Pacific, erupts.
c.1450	Norse settlements in Greenland abandoned.
1453	Eastern Roman Empire falls to the Ottoman Turks.
1450–1530	**Spörer Minimum brings cold.**
1431	Angkor (Khmer) civilization dissolved.
1321–1361	Great Famine in Europe kills tens of thousands.
c. 1250	**Little Ice Age begins.**
c. 1200	**Medieval Climate Anomaly fades out.**
1341–1351	The Black Death kills millions.
c. 1300–1450	Great Zimbabwe in Southern Africa is at the height of its powers.
1220–1448	Mayapán kingdom dominates the northern Yucatán.
c. 1220–1310	Mapungubwe kingdom flourishes in Southern Africa.
1113–1150	Angkor Wat constructed in Southeast Asia, thereafter Angkor Thom.
c. 1050–1300	Cahokia in the Mississippi valley becomes a major political and ritual center.
c. 1017	Anuradhapura kingdom in Sri Lanka disintegrates.

c. 1000	Norse base at L'Anse aux Meadows founded in Newfoundland.
c. 950	**Medieval Climate Anomaly begins.**
c. 850–1471	Chimor flourishes in Peru's north coast.
800–1130	Chaco Canyon becomes a major ceremonial center in the American Southwest.
750–950	At least eight massive volcanic eruptions affect regional climates.
c. 550–1000	Tiwanaku state dominates the highland Andes.
541	Justinian Plague arrives in Egypt and spreads through the Roman Empire.
536	Major volcanic eruption in Iceland.
450–c.700	**Late Antique Little Ice Age.**
405–410	Western Roman Empire dissolves.
330	Constantinople founded as a Roman capital.
166	Antonine Plague in Rome.
c. 100	First subsistence farmers arrive south of the Zambezi River in Africa.
100–800	Moche State thrives on Peru's north coast.

BCE

30	Octavian annexes Egypt for Rome.
c. 200–150 CE	**Roman Climatic Optimum.**
250–c. 900 CE	Classic Maya civilization flourishes in Central America.
377	Anuradhapura, Sri Lanka, founded.
912–610	Neo-Assyrian civilization dominates much of Southwest Asia.
1000–400	Maya farmers move onto the Yucatán lowlands.

147 ½	Queen Hatshepsut's expedition voyages to Punt.
2200–1900	**The 4.2 ka Event (megadrought).**
	Drought throughout Mesopotamia.
	Destabilization of Egyptian state until 2060.
c. 2500	Construction of the Pyramids of Giza, Egypt.
2600–1700	Indus cities flourish.
2334–2218	Akkadian civilization dominates Mesopotamia.
c. 2900–2300	Sumerian civilization in Mesopotamia.
3100	Unification of Egypt brings Upper and Lower Egypt together.
3000–1800	Caral flourishes on the Peruvian coast.
3500	Uruk in Mesopotamia achieves prominence.
6200–5800	Major droughts affect much of the Middle East.
c. 6200	Doggerland finally inundated.
7400–5700	Çatalhöyük, Turkey, a major farming and trading community, comes to prominence.
c. 9000	Farming and herding begins at Abu Hureyra and elsewhere in the northern Middle East.
c. 11,000	Abu Hureyra, Syria, settled by hunter-gatherers.
c. 11,650 (13,650 years ago)	**The Holocene begins.**
c. 13,000 (15,000 years ago)	First settlement of the Americas takes hold.
c. 18,000 (20,000 years ago)	**Natural global warming begins.**

Prolegomenon

BEFORE WE BEGIN

Fires, Ice Ages, and More

As we write this book, fast-moving wildfires have exploded over much of California. Over 1.6 million hectares are in flames so far as dozens of conflagrations, large and small, burn out of control, sometimes joining forces to create much larger conflagrations. Dense ash clouds drift over enormous distances causing serious, health-threatening air pollution. Fire behavior no longer dies down overnight, owing to higher temperatures. The North Complex Fire in Northern California expanded by 40,468 hectares *overnight.* California's annual burned area has increased fivefold since 1972. More than 14,000 firefighters from across the United States and the world have battled the flames. Thousands of people have evacuated as hundreds of houses have gone up in flames. Temperatures have risen; rainfall has diminished and become less predictable; vegetation in often virtually inaccessible terrain is ever drier; mountain snowpacks are evaporating. No less than 30 percent of the state's population lives in areas with wildfire potential, in part because of inadequate land-use policies that encourage urban sprawl. More and more people are building and rebuilding houses in high-risk fire areas. Efforts to manage forests have declined as they are replanted with single species. Almost nothing has been done to encourage people to move out of harm's way. Californians and Oregonians face what appears to be the result of increasingly volatile, catastrophic, and human-caused climatic change: seemingly uncontrollable fire.

This is not the first time in history that people have faced environmental catastrophe, be it flood, drought, or raging flames. But this time it's different. This time, the climate-caused disasters are the direct result of our own recent activities. There are those who are asking whether we will ever adapt to the new realities of extreme temperatures and destructive flames. Our densely populated landscapes are highly vulnerable to the ravages of fires started by lightning strikes and violent downslope winds that carry sparks for kilometers and set entire communities alight in a few minutes. Are we fated for extinction and forced evacuation to safer environments? Or will we adapt to the new, more hazardous conditions that are, to a great extent, of our own making? We are only now confronting these questions seriously.

This book is about human adaptations to climate changes of all kinds. Ancient societies adapted successfully to sudden, short events like ash clouds from distant volcanic eruptions or droughts that persisted for a few years. Our ancestors also adjusted to longer-term climatic fluctuations like rising sea levels, centuries-long arid cycles, and multiyear intervals of much lower temperatures. On the whole, our abilities to cooperate and to provide mutual assistance and effective risk management have served us well. The price has often been high, but the record of history proclaims loudly that we will survive this latest environmental catastrophe. We will achieve this, sometimes at high cost, both by short-term adaptation and, in the longer term, with laboriously debated and implemented permanent changes to society and the ways in which we live.

Fortunately, the past half century has witnessed a revolution in paleoclimatology, the study of ancient climate. What began as bold pioneering efforts in the late nineteenth and early twentieth centuries in the hands of a few gifted scientists has now become major scientific business. The technical literature on ancient climate has mushroomed in recent years. Important papers appear almost weekly, making it nearly impossible even for climatologists themselves to keep up with the literature. Nonclimatologists like us—we are archeologists—sometimes despair at staying current. Even a modest excursion into the academic and sometimes even the more general literature introduces one to a bewildering array of technical terms and acronyms, of which El Niño Southern Oscillation (ENSO) is perhaps the most commonplace.

Our intent in these pages is *not* to delve into the formidable intricacies of global climatology or paleoclimatology, which are a chronicle unto themselves. Rather, we have used the latest information to tell a story about past people and their relationships to the changing climate, from ancient times until the recent past—after all, long chronologies are something at which archeologists excel. As we explored ancient climate change, we discovered that a number of major forces lay behind the climate changes documented in these chapters. These include some familiar phenomena such as El Niños and La Niñas, Ice Ages, megadroughts, and monsoons. This prolegomenon describes these major players in climate change and some others. It also describes some of the "proxies," indirect methods that reveal ancient climate shifts. Think of the rest of this prolegomenon, if you will, as what the great humorist P. G. Wodehouse memorably called "snorts before the solid orgies." If you're unfamiliar with some of these climatic players, indulge yourself with this brief excursion into global climate.

The geoscientist George Philander, an expert on El Niño, sets the stage for us in a classic book about global warming, *Is the Temperature Rising?*[1] He wrote about the asymmetrical coupling of the atmosphere and the ocean, not an ideal pair: "Whereas the atmosphere is quick and agile, and responds nimbly to hints from the ocean, the ocean is ponderous and cumbersome." This one sentence encapsulates one of the fundamental challenges of paleoclimatology: figuring out how an ill-matched pair of climatic giants manage to dance together. Who leads? Who changes the tempo or slows it to a near standstill? Many details of this complex, ever-changing partnership still elude us. We can but examine the major players here.

A PALIMPSEST OF PROXIES

Much global climate change is wrought on a grand scale. Just over a century ago, two Austrian geologists, Albrecht Penck and Eduard Brückner, identified at least four great Ice Age glacial episodes in the Alps, separated by warm interglacial periods. The two geologists worked with glacial deposits in mountain river valleys, but their work is long out-of-date. The four glacial periods are far too simple a portrait of an Ice Age that formed the background to human evolution and the emergence of modern humans on the world stage. Today, we know that the Ice Age (Pleistocene) ended with the Würm glaciation after about 15,000 years ago. With its retreat, the Holocene (Greek: *holos*, "new") brought natural warming and steady progress toward the climatologically modern world.

Our knowledge of Ice Age climate is wrought on a general canvas of cooling and warming. Here, we're working with timescales in millennia and tens of millennia. We know, for example, that the coldest millennia of the last glaciation were around 21,000 years ago. But, as is abundantly clear from later records, climate changes constantly, so a more nuanced portrait of Ice Age climate between 30,000 and 15,000 years ago will come eventually not from glacial deposits but from climatic proxies.

Proxies are sources of climatic information from nature such as glacial cores and tree rings that can be used to estimate changing climatic conditions earlier than the first accurate instrumental records of the mid-nineteenth century. Deep sea cores from the southwestern Pacific extending back 780,000 years, covering much of the Ice Age, reveal at least complete glacial and interglacial cycles during these millennia. Clearly, climate during the Ice Age was far more changeable than had once been suspected. Then came ice cores, drilled deep into the Greenland ice sheet and into Antarctic ice, which are now producing far more accurate records of Pleistocene climate extending back at least 800,000 years. We now know, for example, that a 100,000-year cycle has governed switches from cold glaciations to warmer interglacials over the past 770,000 years. Cooling is gradual; warming is much faster.

There are, of course, numerous complications in the use of deep-sea cores, now available from almost every ocean, and ice cores from many locations, including the topical glaciers of the Peruvian Andes. The proxies from ice and sea cores are becoming ever more precise, but, from the archeological perspective, they generally provide broad climatic background for the Ice Age. So do massive deposits of loess, windblown dust from Ice Age glaciers that often buried late Ice Age settlements in river valleys in the Ukraine and elsewhere. This is admirable as a general perspective, but we have to rely on more fine-grained proxies when considering human adaptations to climatic change.

The word "speleothem" is a tongue twister, and such proxies are relative newcomers on the climatic stage but of great importance. Stalactites (accumulating on cave ceilings) and stalagmites (forming on cave floors) are created by mineral-rich water dripping through the ground into caverns. As the mineral-rich water flows, the speleothems grow in thin, shiny layers. Thicker layers form when more groundwater drips into the cave; thinner ones when less. The layers in a speleothem can be dated by measuring the amount of uranium from the surrounding bedrock that seeps into the water. This forms a carbonate that becomes part of each layer of the growing speleothem. The uranium decays into thorium at a known rate, so the layers can be dated. This creates an approximate record of how groundwater levels changed through time. All kinds of factors affect the growth of speleothems, such as local groundwater chemistry. This means that the climatic record from one cave has to be compared with those from other cave speleothems over a wide area.

Oxygen isotope ratios provide a way of tracking rainfall changes through time, given that both heavy and light oxygen occur in water. Heavy rain provides more light oxygen, and heavy oxygen is a sign of less rain, while different ratios occur in water from different sources. Speleothem research is still in a developmental stage, but it has great potential for providing chronologically accurate rainfall data that can be related directly to past events like the disintegration of Maya lowland civilization in the tenth century CE. Valuable speleothem records are accumulating rapidly in many parts of the world. They are potentially one of the most useful climatic proxies of all.

World geography changed dramatically as sea levels rose some ninety meters to modern levels in a few thousand years at the end of the Ice Age. Two classic examples are described in Chapter 2: the sunken land bridge that once linked northeastern Siberia and Alaska and also the marshy, riverine plains of Doggerland that joined Britain and the Continent until about 5500 BCE. The Sahara Desert sustained cattle herders until about 4000 BCE, during millennia when it supported shallow lakes and semiarid grasslands, known from core borings and pollen analysis.

When we turn to the climate changes of the past 15,000 years, we begin to operate with much more complete proxy data, such as pollen records from North America and northern Europe that document complex vegetational changes as the global climate warmed. The first more precise climatic proxies were tiny fossil pollen grains from northern European bogs and swamps that showed the dramatic vegetational changes after the Ice Age, from open steppe to birch and ultimately mixed oak forest. Pollen sequences, combined with other sources like wood charcoal, now document changing vegetation around early farming villages in western Europe, as well as the cultivation weeds that flourished in cleared fields. For instance, a lakeside settlement in northeastern England occupied between 9000 and 8500 BCE yielded birch pollen and charcoal from reeds burnt repeatedly during autumn and spring when they were dry, when new growth began. This controlled firing encouraged plant growth and attracted feeding animals.

Dendrochronology—dating using ancient tree rings—has been in use for almost a century. First developed in the American Southwest by astronomer Andrew Douglass, who was interested in sunspots, it soon morphed into a method of accurately dating the beams from Ancestral Pueblo ruins such as Pueblo Bonito in Chaco Canyon, New Mexico. Tree rings, formed by the cambium, or growth layer, between the wood and the bark, record the annual growth of certain species of trees, like the Douglas fir in the Southwest. Linked to ring sequences from living trees, ancient rings provide dates for such structures as European cathedrals and Southwest pueblos, shipwrecks, and all manner of other structures. They also provide valuable climatic information acquired by recording oxygen isotope signals created by summer rainfall. Dendrochronology can now be astonishingly precise. Seven thousand tree ring sequences from central Europe have produced estimates of rainfall between April and June, a vital planting and growing season, between 398 BCE and 2000 CE. Tree rings are now big climatological business, with numerous dated sequences from many parts of the world. They not only date archeological sites but provide very accurate charts of wet and dry rainfall cycles. Tree ring sequences are now so plentiful that one can trace the spread of severe droughts across the Southwest. Many of these droughts and other climatic changes resulted from powerful climatic forces with global clout.

THE GULF STREAM

The Atlantic Gulf Stream is part of a vast global conveyer belt of moving water that can alter climate and affect human lives. High-latitude cooling and low-latitude heating—what one can call thermal forcing—drive water flow to the north. Huge amounts of heat flow northward and rise into the artic airmasses over the North Atlantic. The downwelling of salt water in the north nourishes this great ocean conveyor belt that brings warmer temperatures to Europe. This heat accounts for Europe's relatively warm oceanic climate with its moist westerly winds. These have blown, albeit with variations, since the Ice Age.

But not always. As the great northern ice sheets retreated when the Ice Age ended, a huge freshwater lake known as Lake Agassiz, after the well-known nineteenth-century geologist Louis Agassiz, lapped North America's retreating Laurentide ice sheet for 11,000 kilometers. A huge southward bulge of ice blocked the lake water from flowing eastward through what is now the Saint Lawrence valley into the North Atlantic. Inexorable global warming and increasingly little snow accumulation caused the ice to retreat. Then, in about 11,500 BCE, the barrier gave way. An enormous accumulation of glacial meltwater rushed eastward into the Atlantic. The warmer water formed, as it were, a lid on the warm water of the Gulf Stream that flowed north and eastward, keeping Europe warmer. For 1,000 years, the Gulf Stream and Atlantic water circulation ceased. Temperatures in Europe fell rapidly, Scandinavian ice sheets advanced. Europe and the Middle East became much drier. Climatologists call this 1,000-year event the Younger Dryas, named for an arctic tundra wildflower, *Dryas octopetala*, dated with numerous radiocarbon samples to between 11,500 and 10,600 BCE. Then, just as abruptly, Gulf Stream circulation resumed, and gradual warming began, which continues to this day.

The Younger Dryas witnessed major changes in human society, among them the beginnings of farming and animal husbandry in the Middle East (see Chapter 2). Basically modern climatic conditions then came into play. These included irregular, but less prolonged, climatic shifts of much shorter duration. Such changes generated unpredictable rainfall and drought, creating new challenges for human societies. The climatic gyrations occurred during millennia when population densities were rising and settled agriculture became the norm. People had to adapt to them long before anthropogenic global warming came into play.

Rainfall and drought have local impacts, but the climatic forces driving them often originated thousands of kilometers away. The Atlantic Gulf Stream transports warm water from the subtropics to the Arctic. It acts like an air conditioner for Europe, tending to flatten out temperature peaks and troughs. In the long term, modeling suggests there will probably be some weakening of the Stream by the end of the twenty-first century, but much depends on anthropogenic emissions of greenhouse gasses. A worst-case scenario talks about a 30 percent decrease in the circulation, but much depends on the extent to which melting Greenland ice affects it. We do not yet have anything like an accurate forecast.

THE NORTH ATLANTIC OSCILLATION

The major climatic driver for Europe and much of the Mediterranean is the North Atlantic Oscillation (NAO). This giant atmospheric seesaw at sea level between the permanent subtropical high over the Azores and an enduring subpolar low in the north accounts for up to 60 percent of the December–March temperature and rainfall variability across Europe and in the Mediterranean region. It's the dominant agent of winter climatic variability in the North Atlantic and affects a huge area from central North America through Europe and into northern Asia. Unlike El Niños (see below), it is largely an atmospheric phenomenon.

The NAO swings between a positive and negative index. A positive index brings a stronger subtropical high-pressure center and deeper than usual subpolar low centered near Iceland. This means stronger, more frequent winter storms across the Atlantic on a more northerly track. European winters are warm and wet, but the same months are dry in northern Canada and Greenland. The US East Coast winter is mild and wet. A positive NAO index brings cooler and drier winters to much of the Mediterranean and the Middle East. Since the NAO regulates Atlantic heat and moisture funneling into the Mediterranean, both Atlantic and Mediterranean Sea surface temperatures influenced, and still influence, Middle Eastern climate. In general, the influence of the NAO is much less in North America.

A negative index is the opposite: a weak subtropical high and weak subpolar low. The pressure gradient is reduced. Fewer and weaker winter storms cross the Atlantic on a more east-west course. They bring moist air to the Mediterranean and cold air to northern Europe. The US East Coast winter is colder and has more snow. Since the NAO regulates Atlantic heat and moisture funneling into the Mediterranean, both Atlantic and Mediterranean Sea surface temperatures influence Middle Eastern climate. A positive NAO played an important role in providing rain to central and northern Europe during the late third and fourth centuries CE at an important period in the history of the Roman Empire (Chapter 5).

Variations in solar irradiance and cycles of volcanic forcing have accounted for much of temperature variability over the past millennium. This may also have been the case in earlier times, but the NAO is now a major climatic driver over a wide area of the globe. The eastern boundaries of its influence lie in the eastern Mediterranean, at what one can call a climatic crossroads, where Asian monsoon systems and, at a long distance, El Niños in the southwestern Pacific have an effect. This creates wide local variation in both drought and rainfall throughout the Middle East.

MONSOONS

One of our unforgettable experiences was sailing windward in an Indian Ocean dhow against the winter northeasterly monsoon east of Aden at the base of Yemen and at the mouth of the Red Sea. Hour after hour, the lateen-rigged cargo vessel sailed close inshore, shifting onto an offshore tack just outside the breakers. The sea was smooth, the soft tropical wind constant for days on end—or so we were told after a day of memorable passage making. Except for offshore trade wind passages, the Indian Ocean monsoon winds are about as good as sailing gets.

The monsoon world is enormous, extending from Southeast Asia and China through the Indian Ocean, with several major variations on a general timeline of seasonal rainfall. Basically, monsoons are large-scale marine winds, which gain strength when the temperature on land is warmer or cooler than that of the ocean. Land temperatures change more rapidly than those at sea, which tend to remain more stable. During the warmer summer months, both land and sea become warmer, but land temperatures rise more rapidly. The air above the land expands, and an area of low pressure develops. Meanwhile, the ocean remains cooler than the land, with higher pressure aloft. The pressure difference causes monsoon winds to blow inland from the sea, bringing moister air to the land. As the damp air rises, it flows back to the ocean, but the air cools, which reduces its ability to hold moisture, leading often to heavy rainfalls. The opposite occurs during the colder months, when the land cools faster than the ocean and there is higher pressure ashore. The air over land flows toward the ocean and rain falls offshore. The cool air then flows back toward land and the cycle is complete.

For thousands of years, the Indian Ocean monsoon drove navigation under sail. The monsoon trade was lucrative, allowing merchant vessels to complete voyages from India's west coast to the Red Sea or East Africa and back within twelve months. Merchant ships could also sail along the desolate coast between the Gulf and Northwest India. For many centuries, silk and other textiles, as well as Asian exotica, traveled to the west, gold and African ivory to the east. In the Indian Ocean, the southwest summer monsoon blows from July to September, months when moisture-laden air rushes across the hot, arid terrain of the Indian subcontinent. Almost 80 percent of India's rainfall comes from the summer monsoon, with 70 percent of the population depending on agriculture, growing cotton, rice, and coarse grains. Farmers heavily dependent on monsoon rain in western India are extremely vulnerable to delays in its arrival by even a few days or weeks. Numerous famines resulting from monsoon failures killed thousands of people in years like 1877, when the monsoon rains essentially never arrived. More localized versions of the Indian monsoon affect the Arabian Sea and the Bay of Bengal. The southwest monsoon is powerful enough to be felt as far north as Xinjiang in northwestern China.

The East Asian monsoon is a warm, rainy phenomenon that provides an often-wet summer and a cold, dry winter monsoon. Monsoon rains arrive in a concentrated belt, passing north from South China in early May, then to the Yangtze River valley, and finally to northern China and Korea in July. The rain belt moves back to southern China in August. Historically, monsoon rains were of vital importance. The Khmer farmers of Angkor in Cambodia always depended on the Asian monsoon, which is thought to have first developed about ten million years ago, long before humanity appeared on earth. The intensity of the monsoon varied through time, especially soon after the Ice Age, but has always been a dominant presence in global climate. It brings fairly reliable seasonal rainfall as well as arid conditions, or drought instead, to more than 60 percent of the world's population. Differential heating of the Eurasian landmass and adjacent oceans in summer and winter triggers an annual wind reversal on a hemispherical scale. There's another player, too: The Intertropical Convergence Zone (ITCZ), where trade wind belts converge. Three regional monsoon systems also form part of the complex climatic dynamics that affect Southeast Asia. In addition, El Niño events and the Interdecadal Pacific Oscillation create short or longer-term disturbances, which can deliver severe droughts to much of Asia, including Angkor.

The Intertropical Convergence Zone circles the earth at or near the equator, the belt where the trade winds of the Northern and Southern Hemispheres meet. Intense sun and warm water heat the ITCZ's air, raising its humidity and making it buoyant. The buoyant air rises as the trade winds converge; the rising, expanding, and cooling air releases moisture in frequent irregular thunderstorms. Near the surface, winds are normally low, which is why mariners call the ITCZ the Doldrums. Seasonal shifts in the ITCZ affect rainfall in many tropical nations and cause the wet and dry seasons of the tropics. During the northern summer, the ITCZ shifts between ten and fifteen degrees north. This seasonal movement had a powerful effect on rainfall in the Maya lowlands of Central America (Chapter 6). The ITCZ moves northward in the Pacific as the Asian continent is warmed more than the ocean. The warm continental air rises, and air is drawn from the ocean to the land, with southerly winds bringing monsoon rain. The zone then shifts southward during the Southern Hemisphere summer.

EL NIÑO SOUTHERN OSCILLATION

ENSO is arguably the most powerful actor in global climate. Originally, El Niños were thought to be a local phenomenon, which affected the anchovy fisheries off coastal Peru regularly, usually appearing around Christmastime. One of the great triumphs of meteorology lay in the hands of a British Raj meteorologist in India, Gilbert Walker, a statistician by training, who searched for the causes of monsoons and became an expert on El Niños. Walker was one of the first observers to realize that ENSO was a global phenomenon. He concluded that when atmospheric pressure was high in the Pacific region, it tended to be low in the Indian Ocean from Africa to Australia. He called this the "Southern Oscillation," whose gyration changed rainfall patterns and wind directions over both the tropical Pacific and Indian Oceans. Unfortunately, Walker lacked the sea surface and subsurface temperature data to confirm the mechanisms of the Southern Oscillation, which were not available during the 1920s.

A Norwegian meteorologist, Jacob Bjerknes, who worked at the University of California, Los Angeles, also had a global perspective on atmospheric circulation. A strong El Niño in 1957–1958 turned his attention west, which led him to show that there was an intimate relationship in the normal sea surface temperature gradient between the relatively cold eastern equatorial Pacific and a huge pool of warm water in the western Pacific as far west as Indonesia. He argued that there was a huge east-west circulation cell close to the plane of the equator. Dry air sinks gently over the cold eastern Pacific. Then it flows westward along the equator as part of the southeast trade wind system. Atmospheric pressure is higher in the east and lower in the west, which drives this movement. The air returns to the east in the upper atmosphere to complete the circulation pattern. Bjerknes named this the Walker Circulation. He realized that when warming occurred in the eastern Pacific, the sea surface temperature gradient between east and west decreased. This weakened the trade wind flow that drove the lower branch of the Walker Circulation. The pressure changes between the eastern and equatorial Pacific acted like a seesaw, whence the Walker Circulation.

Numerous elements make up the El Niño Southern Oscillation connection. These include the seesaw movements of the oscillation, large-scale air and sea interactions that cause Pacific warming, and much larger global links with climatic changes in both North America and the Atlantic Ocean. Bjerknes showed that ocean circulation is the flywheel that drives a vast climatic engine. Each ENSO has a different personality. Some are immensely strong; others are weak and short-lived, driven in a vast, self-perpetuating cycle between the eastern and southwestern Pacific. Along the equator, a normal north-south circulation known as the Hadley Circulation connects the tropical atmosphere with that of northern latitudes. It carries winter storms northward toward Alaska, except when an El Niño disrupts the pattern. The storm track jogs eastward and hits the California coast.

It took a massive ENSO in 1972–1973 to trigger broad scientific interest in what was now seen as a global phenomenon that upended drought and rainfall patterns with little warning—this quite apart from the collapse of the Peruvian anchovy fishery owing to overfishing. We now know a great deal more about ENSO, a chaotic pendulum with abrupt mood swings that can last months, even decades. The pendulum never follows the same path, even if there is an underlying rhythm to the swings. Tree ring sequences from teak trees in Java, firs in Mexico, and bristlecone pines in the American Southwest, among others, document years of higher rainfall about every 7.5 years until 1880. Now they seem to occur every 4.9 years, with La Niñas every 4.2 years. Ocean corals and ice cores from mountain glaciers tell us that ENSOs have been a factor in global climate for at least 5,000 years, probably much longer. The ENSO cycle is such a powerful engine of global climate change that many experts believe it is second only to the seasons as a cause of climatic change.

ENSO is a tropical phenomenon that has exercised a powerful influence over the lives of millions of tropical foragers and subsistence farmers, as well as preindustrial civilizations in river valleys, in rain forests, and high in the Andes. These societies have always been vulnerable to drought and flood in a world where more than 75 percent of the global population lives in the tropics, two-thirds of which depend on agriculture for their livelihoods. The vulnerability of these societies is increasing daily as growing populations place more stress on the carrying capacity of tropical environments. Until recently, no human society had the ability to forecast ENSOs or other major climatic events. Today, our computers and forecasting models can predict them well in advance. This information is of priceless economic, political, and social value as we adapt to a warming world. None of the ancient societies described in these pages had this luxury, which made adapting to ENSOs an enormous and sometimes deadly challenge.

FINALLY, MEGADROUGHTS

Centuries of tree ring research have provided ample records of prolonged megadroughts—the term is used in the climatological literature—during the Medieval Climate Anomaly (c. 800 to 1300 CE) and the Little Ice Age (c. 1300 to 1850 CE), both of which we explore in Chapters 11 to 14.

Tree ring–based climate sequences have the advantage of being very accurate, to the year. Fortunately, tree ring sequences are now abundant over a wide area of the United States, so much so that the *North American Drought Atlas*, compiled by an impressive team of climatologists, reconstructs 2,000 years of summer moisture, using a yardstick known as the Palmer Drought Severity Index. The latest version of the *Atlas* highlights two megadroughts that had major impacts on Native American societies. One, in the Southwest during the late 1200s, contributed to the depopulation of the Mesa Verde and Four Corners regions by Ancestral Pueblo groups (see Chapter 8). The second occurred in the Central Plains during the 1300s. The megadrought immediately preceded, and continued after, the abandonment of the great ceremonial center at Cahokia, in Mississippi's American Bottom (also see Chapter 8). Understanding of the impacts of these and other droughts is hampered by the uneven distribution of tree ring sequences, especially in areas like the Central Plains.

Recent megadroughts in the American West have been severe, but they were much more prolonged over the past two millennia. They were certainly much longer than the celebrated Dust Bowl drought of 1932 to 1939. A series of influential studies have shown that megadroughts affected nearly every area of the West but also occurred during the early to middle Common Era in Mexico, the Great Lakes region, and the Pacific Northwest.

Until the mid- to late nineteenth century, ancient societies had to adapt to climatic shifts that were of natural origin, most of them attributable to the powerful forces that dominated global climate change in the past. As fossil fuels and intensifying industrial activity took hold, so anthropogenic forcing (changes in the earth's energy balance due to human economic activities) came into play, and our current climate crisis took hold. But to combat its potential devastation, it's important than we understand the forces behind natural climate change over the centuries and millennia.

1

A FROZEN WORLD

(c. 30,000 to c. 15,000 Years Ago)

Central Europe, autumn, 24,000 years ago. Two weather-beaten hunters sit on a streamside boulder with their backs to the wind, their faces turned toward the horizon. They ignore the reindeer feeding on the other bank, pawing through the dead leaves of autumn. Gray clouds scud close to the ground, massing heavily to the north. Neither of them says anything as they watch the cold, arid landscape in the growing darkness. Then they look at one another and nod, gathering their parkas close around their shoulders.

Their summer dwelling hugs the ground, a low dome of sod and hides. The hunters stoop into the smoky interior, where everyone clusters around a blazing hearth as fat lamps flicker in the gloom. As darkness falls and the churning storm strengthens outside, the people huddle under furs and hides. One of the hunters, believed to have supernatural powers, tells a familiar tale of mythic beings, of the first people of long ago. The band has heard the story many times, a tale of constant movement following the reindeer and wild horses in spring and autumn. As the story unfolds, the elders listen to the opinion of everyone, young and old, male and female. It is time to move to winter quarters.

We are *Homo sapiens*, the self-named "wise people." Our species emerged in the warmth of Africa at least 300,000 years ago—the date is still controversial. Nimble, intelligent creatures, we moved over wide hunting territories and adapted to climatic shifts like long drought cycles by anchoring ourselves to reliable water sources. We were consummate opportunists, relying on careful observation, an intimate knowledge of the surrounding landscape, and cooperation—both within the narrow boundaries of family and band and also with extended kin—for survival. We lived with the vicissitudes of local climates using simple, portable tools and weapons. For almost all of our existence, we led this nomadic life—moving with the animals and with the seasons. Before writing, which emerged in western Asia a mere 5,000 years ago, we passed on all knowledge—real or imagined—by word of mouth and sometimes via art as well.

Our survival in ancient times depended on intimate knowledge of, and respect for, the living world of which we are part. Though no present-day group is a portal to the remote past, it is helpful to consider what life is like among today's few remaining nomadic hunter-gatherer societies. Among the Inuit of the Arctic or the San of Southern Africa, we find a strong deference for prey and a deep understanding of the living environment—of the seasons, of plant foods, and of the migration of game animals. This knowledge spells the difference between life and death, as it always has.

Back in our African homeland, violent storms, intense drought cycles, and the ash-laden aftermaths of major volcanic eruptions were always climatic realities. But the challenges to our survival intensified dramatically when some of *Homo sapiens* moved into the much colder, sparsely inhabited European and Asian world around 45,000 years ago. We found ourselves battling some of the harshest climatic environments our species has ever lived through. But there was more: we were not alone. During the six-million-plus years of human evolution, several distinct hominin species always coexisted at any given time.

For example, in Eurasia, there were Neanderthals between about 400,000 and 30,000 years ago, though the dates are disputed. These were hominins with whom we were reasonably closely related in evolutionary terms. We shared a common ancestor back in Africa, up to if not before, 700,000 years ago. On an island in Southeast Asia until about 50,000 years ago, there existed an isolated population of another diminutive human, *Homo floresiensis*, the so-called hobbit folk, known for their short stature. A third, still largely unknown hominin species—the Denisovans—lived in Siberia and further east and south.

There were other hominin species, too, of which we know almost nothing. And despite some minimal interbreeding between some of the various species (notably *Homo sapiens*, Neanderthals, and Denisovans), all, bar us, were destined for extinction. By 30,000 years ago, we *Homo sapiens* were the only hominin species to remain standing.

Exactly why all the other hominin species became extinct remains one of the enduring debates surrounding prehistoric times. Their disappearance tends to broadly correspond with the arrival of *Homo sapiens* in each region. This has led many to argue for a "them versus us" situation in which we killed them off, outcompeted them, or both. However, it is equally feasible that the different species had relatively little to do with one another—beyond the occasional, sometimes amorous, encounter. Perhaps something else more serious was afoot. Evolutionary geneticists, such as David Reich, argue that the Neanderthals had already been in decline since about 100,000 years ago, potentially as a result of harshly changing climates, and only a few thousand remained by the time *Homo sapiens* arrived in their homelands. Similar environmental stories may help to explain the demise of some of the other now extinct hominins. The only certainty is that, ultimately, *Homo sapiens* settled throughout the world, adapting to new and sometimes challenging environments.

A DIFFERENT WORLD

What was this past world like? The world of 45,000 years ago was unimaginably different from the warming globe that supports over 7.5 billion people today.[1] Vast ice sheets covered northern Europe and flowed out from the Alps. Two great ice sheets extended deep into North America as far south as Seattle and the Great Lakes. Africa's Kilimanjaro and Ruwenzori mountains, South America's Andes, and New Zealand's Southern Alps were icebound, all in addition to the great deep freeze in Antarctica. With so much water absorbed into ice sheets, global sea levels were around ninety meters (three hundred feet), or more, lower than today. You could walk dry shod from Siberia to Alaska across an intensely cold and windy land bridge. The North and Baltic Seas were dry land; Britain was attached to the European continent. Huge coastal plains extended out from the

Southeast Asian mainland toward New Guinea and Australia. Vast tracts of scrub-covered arctic tundra ranged from the Atlantic coast deep into Europe and Siberia. For months on end, boisterous northerly winds mantled the endless, dry steppes in fine glacial dust from the northern ice sheets. Over much of Europe and Eurasia, animals and humans endured nine-month winters and persistent subzero temperatures. How cold was it? Climatologist Jessica Tierney and her colleagues have developed models that use data from ocean plankton fossils combined with climate simulations of the last glacial maximum to reconstruct sea surface temperatures. Their research has established that average global temperatures were 6°C lower than today, with the greatest cooling in high latitudes, as one might expect.[2]

Data from Greenland ice cores have shown that those who lived in the north adapted to an intensely cold, climatically fickle world during a period of constant, sometimes rapid climate change. Globally, temperatures dropped much more in the Northern Hemisphere. This was mainly because of the moderating effects of the oceans. Water covers 60 percent of the earth's surface in the Northern Hemisphere and closer to 80 percent south of the equator. This means that temperatures tend to be warmer on land in the Southern Hemisphere, except, of course, near Antarctica. Winter temperatures are lower in the north, with much greater seasonal differences and more intense cooling further from the equator. About 24,000 years ago, temperatures near New York dropped by as much as 10°C and in the Chicago region by 20°C. In contrast, the temperature drop in the Caribbean area was only about 2°C. Steeper temperature gradients between the North Pole and the equator generated significantly greater wind velocities, which caused windchill factors to climb to dangerous levels for both animals and humans.

Yet there wasn't a permanent deep freeze. Greenland's ice cores reveal more than twelve short warmer periods, known as Dansgaard-Oeschger events, between 60,000 and 30,000 years ago. A sudden warming episode in Greenland 38,000 years ago caused average temperatures to jump by 12°C within a remarkably short time, perhaps as little as a century. Local annual temperatures probably remained between 5°C and 6°C below today's levels. Equally brief, cold intervals brought plummeting temperatures that were between 5°C and 8°C below those of the warmer oscillations.

Among *Homo sapiens*, around 35,000 years ago, our population growth seems to have slowed in the north, perhaps due to expanding ice sheets.[3] The numbers of people may have actually declined as small family bands retreated into more sheltered locations like deep river valleys and mountain valleys, closer to the Mediterranean. Only a few hundred hunting bands dwelled in Europe. During a life span of about twenty to thirty years one probably met no more than a few dozen individuals, many of them living in other groups. Without such contacts, no one would have survived, for no hunting band, however expert, could be completely self-sufficient, especially in daunting Ice Age landscapes. From the beginning, our forebears relied heavily on ties of kin for information, expertise, and mates. Mobility and contacts with others spread technological innovations over enormous distances within remarkably short spans of time. Fortunately, during the coldest millennia, the population never dropped enough for people to lose their vital technological adaptations to bitter Ice Age cold or their intricate symbolic relationships with the supernatural world, which helped sustain them.

Full glacial conditions returned after 30,000 years ago, with extremely cold temperatures between 24,000 and 21,000 years ago. These were the coldest millennia of the late Ice Age, commonly known as the last glacial maximum. With water locked away in ice, global sea levels were about ninety-one meters lower than today.

WRAPPED UP

How did late Ice Age people adapt to such extreme cold? We *Homo sapiens* are all (essentially) hairless apes of African origin. Without clothes, our bodies react to cold when the air temperature drops below 27°C. At 13°C we start to shiver. Yet these are only laboratory readings from people standing in stationary air. An unclad body will lose heat more quickly when the wind is blowing. Even mild subzero temperatures can be hazardous for unclothed people. At –20°C with a thirty-kilometer wind, frostbite takes hold in less than fifteen minutes.[4]

Add moisture to the cold equation, as water on our body surfaces condenses as it cools and sweat becomes a serious issue. The sweat soaks clothing: the clothes lose their thermal and insulating functions. If we get too cold, our core body temperature sinks below the critical level of 37°C. When this core temperature falls due to failed thermoregulation, our bodies enter hypothermia. When body temperature reaches 33°C, we slip into unconsciousness. Below 30°C, the heart slows down, blood pressure drops, and cardiac arrest is almost inevitable.

How, then, did our forebears adapt to the extreme cold and sudden temperature changes of the late Ice Age? Most of us today set our car heaters to around 21°C, a temperature at which we are comfortable—wearing clothes. But we know that people who, from birth, have lived without garments are better able to cope with exposure to cold. When Captain Robert FitzRoy of the HMS *Beagle* surveyed the Magellan Strait in 1829, he encountered the Yaghan people. At the time of his visit, there were perhaps 8,000 of them, short, chunky people who stood, on average, about 1.5 meters tall. Despite the cold temperatures and frequent rain and snow, they usually went naked, in cold weather using otter or sealskin capes that reached to the waist. Young Charles Darwin accompanied the *Beagle* in 1833 and was shocked. "Four or five men suddenly appeared on a cliff; they were absolutely naked & with long streaming hair."[5] Their tolerance for cold was remarkable.

Apart from adaptation in general terms to low temperatures through both mobility and superficial genetic changes, the only weapons that humans possessed against the cold were fire, clothing, and efficient stone tools. No one knows precisely when we first domesticated fire, but our hominin ancestors seem to have been sitting around a (purposefully controlled) fire by around one million years ago, according to recently returned evidence from the Wonderwerk Cave in South Africa. Fire was, of course, unimaginably important. It provided numerous benefits, from protection against wild animals to a boost in the calories available from food due to cooking. (The release of more energy from food for our calorie hungry brains may have also been a critical driver in human evolution.) But arguably of highest importance, fire helped keep us warm. It allowed people to survive in colder environments as they expanded out of Africa and also eased cold nighttime temperatures for those who remained in the homeland. Fires were even used to clear out caves before people lived in them. It is also worth remembering that cave dwelling was itself a development that meant protection not only from predators but also the weather.

As to clothing, the other great protector against the cold, the basic principle is simple: cover yourself. As with many other innovations, the notion of draping hides and other coverings over one's body took hold on many occasions and at different times with no precise date of invention. In the simplest iteration, one simply draped one's upper body in skins, like the Fuegian people, the San hunters in Southern Africa's Kalahari Desert, and the Australian Aborigines. More than mere garments, these skins served many purposes: carrying young children when slung over a shoulder, transporting nuts and other plant foods back to camp, protecting one's hands during stone tool knapping, or carrying freshly butchered meat. People slept in cloaks and buried the dead in them.

Caribou or reindeer hides sufficed in colder climates; the San wore antelope skins in the tropics. The Hawaiians and New Zealand's Maori created feather cloaks, which were garments of high prestige. A cloak could be slipped on or off in seconds and was never fitted tightly. They were highly practical, versatile drapes that became much more effective in lower temperatures when wrapped closely around the body.

No one could survive a northern winter during the Ice Age without thick garments. Neanderthal sites from the bitterly cold millennia of the last glacial period 50,000 to 60,000 years ago yield large numbers of stone scrapers with long, carefully shaped edges used to process skins for bedding, cloaks, and other purposes. But the technology of our Neanderthal cousins was not adaptable enough to do more than prepare a draped hide—although perhaps they sewed using sharpened rocks or thorns as needles, which would be long lost to the ravages of time. To be fair, finding needles in the archeological record is worse than searching through a haystack. However, a team working in the *Homo sapiens* cave site of Sibudu in South Africa did manage just that when they found a 61,000-year-old point that may be part of a specially made bone needle.

Certainly, at some point during the last Ice Age, people living in Europe and Eurasia realized that extra layers of more closely fitted clothing added to personal insulation and kept them warm even in very cold conditions. But to be truly effective, the inner layers had to be tailored to individual limbs, to hips and shoulders, using thread made from animal sinew or plant fiber. The eyed bone needle and the finely crafted awl made tailoring possible. As ever, necessity was the mother of invention, and examples come from Europe and Siberia. However, the very oldest known example, a bird bone needle dated to around 50,000 years ago, comes from the Denisova Cave in Siberia and is attributed not to *Homo sapiens* but to the Denisovans—the species of human who had lived in Europe for millennia before our arrival. These tools point to hominin ingenuity and fueled the first technological adaptations to unpredictable climate: versatile clothing for various temperatures. Yet all these other hominin species perished, for unknown reasons, although climate change may well have been at work. After 30,000 years ago, only one human species still walked the earth: us, *Homo sapiens*.

CUTTING-EDGE TECH

Despite the often unremitting cold, *Homo sapiens* flourished in their new European homelands. They may have arrived from Africa during a warmer interval, which resulted in slowly rising population densities and important changes in hunting weapons. These innovations came about not necessarily as a result of deliberate migrations but through regular contact and exchange of ideas among people in different bands, as South African archeologist Lyn Wadley has pointed out. We know that technological changes happened in Southern Africa at least 70,000 years ago, when small, lethally sharp stone spear points came into widespread use over a large area. We can be sure that the same processes with other ideas, other technologies, took hold in the unfamiliar environments north of the Mediterranean.

We do not know when members of our species moved northward into Europe and Eurasia, but, most likely, they first migrated onto the

East European Plain north of what is now the Black Sea, then a large glacial lake, about 45,000 years ago.[6] Far to the west, there was potential competition. Neanderthal bands had adapted successfully to relatively cold conditions long before we arrived on the scene. This may be why *Homo sapiens* first settled in the colder and less inviting east, visiting seasonal camp sites above the Arctic Circle at sixty-six degrees north. But by 35,000 years ago, *Homo sapiens* were well established in some numbers in the heart of Neanderthal lands in the west, although some moderns must have been there as much as 10,000 years earlier.

It is striking how *Homo sapiens* adapted so rapidly to such a broad range of environments in a short time. As they dispersed widely over Europe and Eurasia, they brought with them elaborate symbols, beliefs, and concepts of time and space that formed distinctive worldviews and behaviors. A critical element was fluent speech and language, which, while probably not exclusive to our species, certainly enabled our forebears to construct their worlds through words and sentences and through art. They interpreted the environments around them—animals, clouds, cold and heat, snow, rain, and drought—using chants, dances, music, and song. Sound-producing drums and other musical instruments, such as flutes, however, rarely survive in archeological sites. These were the devices that people used to structure their cosmos and surrounding world, in both pragmatic and symbolic ways. This reflects more closely organized settlements than those of other now extinct human species. Changing tree colors, the passage of the seasons, the cycles of heavenly bodies: these and other forms of symbolism, such as changing cloud formations, measured the passage of time and the realities of space.

Like modern-day arctic peoples and hunters and foragers everywhere, the new inhabitants of northern lands compiled vast inventories of the environments around them. Their knowledge of the uses of plants alone would have been encyclopedic, as it would have been of ice and snow, with different terms for their distinctive characteristics. This knowledge passed from generation to generation—everything from the best materials for fashioning a ptarmigan trap to the proper curing of skins for hooded parkas.

Nearly all of this knowledge was intangible, unwritten, and ephemeral. As archeologists, we can only speculate intelligently about the means by which the remarkable and increasingly complex technology used by *Homo sapiens* in the north came into being. The technology first appears in tropical Africa, where stoneworkers developed ways of producing small, razor-sharp tools. This in turn led to more sophisticated ways of making tools from various raw materials, including antler, bone, shell, and wood.

A family on the move could chip off a few narrow blades from, say, a flint nodule in seconds and then turn them into different relatively specialized tools. They made everything from razor-sharp spear points to scrapers, awls for punching holes in leather and wood, and chisel-like devices known to archeologists as burins. Such easily portable tools were sharp enough to cut long strips in antlers, which could then be fashioned into harpoon heads and other weapons. These remarkable toolmakers used a carefully trimmed lump of fine-grained stone as the matrix for creating specialized tools. The technology mastered in the north has remarkable modern analogies in the Leatherman or Swiss Army knife. Among a dazzling array of implements was, arguably, one of the most useful and enduring tools ever devised by humanity: the eyed needle.

The needle, the pointed stone awl that pierced hide, a sharp knife edge, and fine thread cut from sinew or vegetable fiber: these humble tools revolutionized life in severely cold landscapes.

SHARP DRESSING

Clothing is perishable and rarely survives in archeological sites. As when studying climate change, one has to rely on proxies, in this case the telltale wear of the most prosaic of tools such as stone knives and scrapers. For instance, the edge wear on blades and scrapers from the Pavlov settlement in the Czech Republic, occupied between 22,000 and 23,000 years ago, reveals edgewise use of stone blades—as if habitually for cutting. They enabled people to create complex garments fitted to shape that covered the vulnerable torso and encased limbs like cylinders. The tailor could not only create complex clothing but use carefully selected materials for different parts of, say, a parka and its hood or for shoes, taking advantage of the unique hides and skins from such animals as reindeer and arctic foxes. Three or four layers, from underwear to water- and wind-proof anoraks and trousers, allowed people to function efficiently in subzero temperatures. All these garments were carefully fitted. One can make very adequate fitted clothing using fine awls to pierce holes, which was likely the tool of choice when moderns first settled in the colder, northern latitudes. But the eyed needle allowed for far more intricate clothing such as underwear. Layered clothing had the advantage that one could easily don or take off surplus layers if temperatures changed rapidly.

Wearing fitted garments meant that people became acclimated to clothing, which made coping with cold without it harder. More complex, layered clothing mediated the shock of walking from a warm rock shelter into the frigid open air. As any runner or cyclist will testify, quickly adding a layer of clothing protects against rapidly changing temperatures, rain, snow, or wind. All modern-day protective clothing works on the principle of layers.

Simple clothing developed into more elaborate fitted garments as conditions became colder and more challenging. At Shuidonggou in northwestern China, eyed needles came into use around 30,000 years ago as the climate grew colder at latitudes between thirty-six and forty degrees north.[7] Far to the west, eyed needles appear around 35,000 years ago in the Ukraine at fifty-one degrees north. They were in use in western Europe, where temperatures were somewhat milder, by around 30,000. Eyed needles became more commonplace during the coldest period of the last glacial maximum around 21,000 years ago.

Fitted clothing and the technology used to fashion it, along with an intimate knowledge of the environment and unceasing mobility, were the primary ways in which humans adapted to the constant and sometimes rapid climate changes of the late Ice Age, specifically the cold.

COLD COMFORT

For all their magnificent art on antler, bone, and cave walls, we have no self-portraits of hunters among the animals they depicted so brilliantly as early as 35,000 years ago (again the dates are disputed). Indeed, we find only the rarest glimpses of the people, such as an approximately 25,000-year-old ivory-carved head from southwestern

France, dubbed the "Lady of Brassempouy." The lady (though it looks more like a young girl or perhaps a boy) is Europe's oldest-known realistic representation of a human face. Its meaning has been endlessly debated, but the head is covered with an angled pattern that falls over the shoulders, variously interpreted as a wig or a hood. More likely, the pattern is simply the person's tightly braided hair. This hairstyle is not surprising, as the current genetic evidence shows that, at this time, Europe's *Homo sapiens* had curly hair and black/dark skin, a clear reminder of our African origin.

Many of these early hunting bands dwelled in rock shelters, under large and small natural overhangs that lined the sides of deep river valleys. Great shelters like La Ferrassie and the Abri Pataud near the village of Les Eyzies in southwestern France were occupied, at least seasonally, over long periods. There are signs that the inhabitants of

The Lady of Brassempouy, France, an ivory figure excavated in 1894. *Natural History Museum/ Alamy Stock Photo.*

these and other shelters hung large hides from the overhangs to create warmer dwellings sheltered from piercing winds, with large hearths and sleeping places behind them.

The busiest times of year must have been spring and autumn, when bands came together to hunt the migrating reindeer herds. The autumn migration was important as the beasts were fat after the warmer months. This was when the people acquired their stocks of hides, fat, and dried flesh to meet winter needs. Thanks to studies of ancient and modern reindeer teeth, we know of eight reindeer ranges of two hundred to four hundred kilometers, three of them in southwestern France, where *Homo sapiens* also lived.

The packed occupation layers of the Abri Pataud rock shelter on the Vézère River tell us that life changed little between 28,000 and 20,500 years ago, a period of intense cold.[8] About 24,000 years ago, the occupation centered around a substantial tentlike structure erected between the cliff and some boulders at the front of the shelter. Poles covered with sewn hides between the back wall and the floor formed a sturdy dwelling. One can imagine smoke from the hearths hovering under the overhang on windless days. The inhabitants hunted wild horses, reindeer, and the fierce aurochs—a large wild ox that survived in Europe for thousands of years before going extinct in 1627. By any standard, these early people were efficient, ingenious hunters who knew their demanding environment and its many climatic vicissitudes in ways that are unimaginable today. You only have to look at their magnificent depictions of bison and other animals to realize that they spent a great deal of time observing the specific habits of their prey. Images of reindeer mating, the summer and winter coats of horses, a bison grooming its flank, or animals alert or in threat postures reveal a deep understanding of their living environment.

Above all, the art of late Ice Age humans reveals a complex relationship with the natural world and the supernatural forces of the cosmos around them. Hand imprints persist at many sites, as if the visitors acquired some form of power from contact with the painted rock faces far below the earth's surface. Two black horses in France's Pech Merle cave, painted about 24,600 years ago, face one another, surrounded by large black dots and stenciled hands. At Gargas cave in the foothills of the Pyrenees, generations of men, women, and children, even infants, left their hand impressions on the walls of the cave's lower level. We can only imagine their meaning. Were the prints perhaps protective marks (touching stones for luck is a common motif among many humans), providing indelible proof of contact with the supernatural realm? They may have had other functions, too, perhaps serving as a means of marking human relationships, stating one's affiliation to the group, and so on.[9]

Despite early humans' understanding of the environment and food sources, climatic shifts were out of their control and never ceased. Years of prolonged cold and scarcity were followed by warmer cycles with plentiful game. Like all hunters and foragers, late Ice Age people took advantage of every chance to kill their prey in strategic locations. About 32,000 years ago, at Solutré, near the modern city of Mâcon in central France, late Ice Age hunters engaged in an unchanging routine of slaughter year after year in a natural corral.[10] During the cold years of the last glacial maximum, wild horses and reindeer abounded on the surrounding open steppe. Between May and November, the hunters would lure young stallions into this corral and slaughter them in a frenzy of killing and butchery. At least 30,000 horses perished in the Solutré hunts over the millennia; decaying carcasses and skeletons paved the valley. The hunts continued until the intense cold around 21,500 drove the hunters southward into warmer environments.

Far to the east, among open plains, sheltered valleys, and mountain foothills, hunting bands with different traditions met, mingled, and maintained contacts. They were able to interact with people living far away on the margins of the great steppes to the east and in the shallow river valleys that extended to the Ural Mountains. The lands in the east were a brutal, cold world of gray-brown dust and wind, with unrelenting aridity. Harsh as they may have been, the eastern plains supported a surprisingly rich bestiary and tough hunting bands that preyed upon them.

Some of the best-preserved camps lie along the Don River, where people hunted mainly horses and fur-bearing animals before about 25,000 years ago. They briefly occupied open-air encampments during the summers, when widely dispersed groups came together to trade, arrange marriages, settle disputes, and perform ceremonial observances. Such open-air encampments were abandoned during the long winter

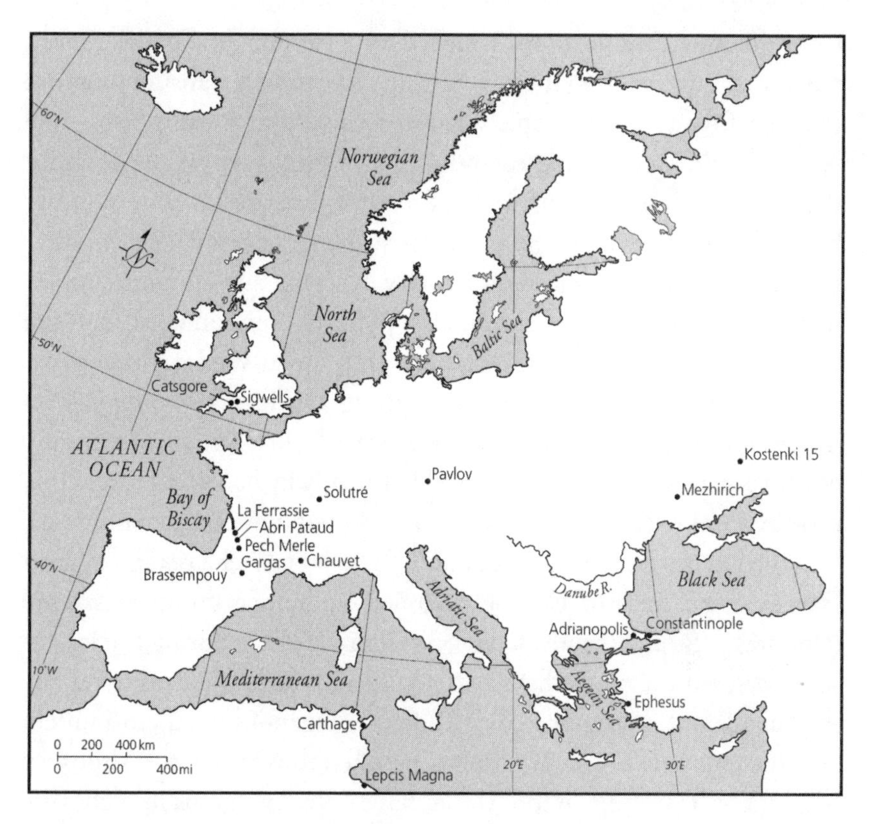

Locations in Europe mentioned in Chapters 1 and 2.

months, when the people dispersed into smaller bands and occupied semisubterranean dwellings dug into the underlying glacial dust.

The Mezhirich site in the Dnieper River valley in Ukraine dates to about 15,200 years ago, long after the last glacial maximum. Temperatures may have been somewhat warmer, but winters were still intensely cold.[11] The people adapted brilliantly by moving partially underground into dome-shaped dwellings about five meters across. They fabricated the outer dome-shaped retaining walls from carefully patterned mammoth skulls and bones, forming the roofs from hides and sod. American archeologist Olga Soffer has estimated that about fourteen or fifteen workers would have built four houses clustered at Mezhirich in about ten days.

Places like Mezhirich were base camps, used for as long as six months a year, built in shallow river valleys that provided some shelter from the relentless northern winds. Come summer, the bands moved out into more open country, where they occupied transitory camps. Perhaps fifty to sixty people occupied each winter camp, with one or two families occupying each dwelling. During the winter months, they would have lived off carefully stockpiled meat from summer hunts, buried in pits in the permafrost. Here, as elsewhere, hunting migrating reindeer was a staple activity in spring and autumn, supplemented by trapping smaller animals and birds, even some fish. But the greatest emphasis was on fur-bearing prey, for survival in these harsh environments depended on fitted clothing and furs. Trapping was an important skill in subzero climates, carried out with simple, highly effective snares set along paths used habitually by hares and foxes. This provided not only food to survive the cold but materials to endure it.

During the last glacial maximum, there was never a perennial deep freeze in the areas of Europe occupied by human groups. As temperatures rose, bands moved out of sheltered valleys during the longer summers, but they must have had no illusions as to the dangers of being caught out in extreme cold. Today's indigenous northern hunters are reluctant to take long hunting trips when the temperature drops far below zero. Traveling on foot would have been dangerous in such conditions, as the people of the day must have known. Some folk hunted, while others stayed in camp, where they would have spent much of their time making clothing and preparing hides and furs—scraping removed fat to make them supple. Late Ice Age people tanned a wide range of skins, even those of birds, carefully scraped and treated with oil. The endless scrape of stone against hide was the late Ice Age equivalent of today's city traffic sounds—an unceasing necessity.

An intimate knowledge of the changing environment, carefully timed mobility, and a deep respect for the natural world: these were humanity's essential adaptive skills in this primordial Ice Age world, passed from generation to generation, as the years, centuries, and millennia unfolded. Above all, as gregarious pack animals, our early ancestors depended on carefully nurtured community ties, constant intelligence acquired from others, and cooperation—cooperation being one of the most enduring of human qualities for adapting to climate change. Cooperation was the glue that held life together. With very small numbers of people living across harsh, predator-rich landscapes, virtually every activity, even preparing skins or sharing the meat from a kill, involved not just individuals but families and the group as a whole. Hunts were rarely solitary, for watchful pairs of hunters had a higher chance of success and more safety. Women often foraged for edible plants and nuts in close-knit groups, sometimes within a few hours' walking distance. Their cooperation made use of intelligence about nut groves and other foods acquired throughout everyone's lives, whether young or old. The unspoken agenda was preventing food shortages, accumulating food to be both eaten at once and also stored away in pits, caves, and rock shelters for consumption during the winter months. There were other realities, too. Hunter-gatherers living in small bands dwelt in temporary camps or more permanent cold weather locations, but numbers were so small that an accident could wipe out two expert hunters in a moment and decimate a band. A sudden freeze could destroy a nut harvest overnight and threaten winter food supplies before they were gathered. Childbirth could kill a mother and leave a helpless orphan. Under such circumstances, people relied on one another to survive, be it on other band members close at hand or others living some distance away.[12] Putting it bluntly, one couldn't function or survive without other people and without the close ritual ties that bonded families, kin, and bands.

In the past, community and kin tied small hunting bands together in tight bonds of cooperation. People survived because of other people, both close at hand and further afield. Cooperation meant success in the hunt and while foraging; the collective expertise of the band ensured survival and lessened risk, reinforced by chanting, singing, and storytelling during summer evenings and winter nights that defined both existence and the fundamental behaviors based on cooperation that ensured survival in good seasons and bad. The hunter's world was vibrant and alive with living forces. It is, and was, no coincidence that people respected their prey and their ever-changing environments. These ancient qualities of cooperation and a carefully nurtured knowledge of the landscape survived in human society for thousands of years and still endure in a few societies today. Unfortunately, many of them have vanished or been undervalued in our crowded, industrial world.

But, as the experience of extreme climatic events, like Hurricane Katrina, has taught us in recent years, we need many of these ancient attributes more than ever. The aftermaths of hurricanes and of the massive wildfires caused by extreme heat, lightning strikes, and downslope winds that have gutted small, rural California communities have brought seemingly anonymous neighborhoods together to cooperate in rescue and recovery. This is when people rely on kin ties and close-knit organizations like church congregations or clubs to provide shelter, food, and support. During such times, the common good becomes more important than individual goals. It is as if cooperation is wired into us. And yet most of us live in conditions so far removed from the frozen world of 20,000 years ago that we fail to look back and learn. Indeed, even our ancestors back then were on the cusp of change—for the Ice Age was about to end, an irregular, then intense global warming was about to begin, and before long the way most of us lived would never be the same again.

2

AFTER THE ICE

(Before 15,000 Years Ago to c. 6000 BCE)

The northern Middle East, today's Lebanon, about 12,000 years ago. Summer days were unseasonably chilly, the sky masked by gray clouds. The band members shivered in their camp among the oak trees. They had settled there some time ago, attracted by the fast-flowing stream that ran nearby. Day by day, the stream dried up until mere trickles became stagnant in ever-smaller pools. The rains were a fraction of those recalled in the elders' memories. Everyone was hungry, subsisting on trapped birds and rodents and wild grasses that clung to life among the trees. Around the campfires, the elders discussed reports of standing water and more plentiful foods in a nearby valley. They listened to young and old alike and decided to move. The next day, the band shouldered its belongings and began a search that, unbeknown to them, would last for generations.

Hunters and foragers had flourished in the relatively well-watered lands between the Mediterranean coast and the Syro-Arabian Desert for millennia. The previously icy climate of the late Ice Age had warmed slowly after 20,000 years ago. By 14,500 to 12,700 years ago, the local people lived in an Eden: warm and moist, with increased rainfall and more predictable food supplies. But now, seven centuries later, something was afoot. Temperatures were rapidly dropping. The future of the band was unclear. How do we know all this? The answers lay in glacial deposits and lake cores deep within Africa's Ruwenzori mountains on the border of today's Uganda and the Democratic

Republic of the Congo. Geologically, these great peaks are one of the places that tell us about ancient climate change, including, critically, when warming at the end of the Ice Age began.

READING ANCIENT CLIMATE

The Ruwenzori mountains (or Rwenjuras, as they are known locally) peak as high as 5,100 meters and have five vegetational zones that alternate from tropical rainforest to alpine meadows and snow-covered terrain.[1] During the last glacial maximum, 24,000 years ago, glaciers from the mountains' central peaks flowed down the valleys that cut through the Ruwenzori. The icy masses merged to form a single superglacier at about 2,300 meters above sea level. The glacial rubble of the long-vanished ice sheet now encloses an entire lagoon, Lake Mahoma, at an altitude of 3,000 meters. Today, lush tropical vegetation covers the valleys. The precipitous mountain slopes tower high above, the snow-capped summits often mantled with cloud. Nonetheless, all is not well: in 1906, the Ruwenzori had forty-three named glaciers distributed over six mountains covering 7.5 square kilometers. In today's warming world, only three mountains still possess glaciers covering a mere 1.5 square kilometers. The long-term thawing has had a massive impact on the Ruwenzori's vegetation and biodiversity.

Modern barometers of climate change, these ice-capped mountains are thawing at an alarming rate. However, they also provide critical information about the ancient past. When glaciers advance, they drag mounds of rock and dirt with them; when they retreat, cosmic rays bombard these newly exposed ridges of rock and dirt (accumulations of glacial rubble known as moraines). By crushing this material and then measuring the buildup of the cosmogenic isotope beryllium-10 (or 10Be), scientists are able to pinpoint when the glacier receded, track its retreat upslope through time, and indirectly calculate how much the climate warmed. Consequently, we know that the Ruwenzori glaciers were at their greatest extent until between about 21,500 and 18,000 years ago and then began their inexorable retreat starting somewhere between about 20,000 and 19,000 years ago, in the face of warmer global temperatures.[2] This was the geological moment when the earth's current natural warming began, heralding the end of the last Ice Age.

Cores bored into East African lakes tell a similar story. By 19,000 years ago, warming of the tropical oceans was well underway. This was also when the great Laurentide ice sheet that mantled the northern latitudes of North America began to retreat, as did ice sheets in the Southern Hemisphere. Epochal changes in the late Ice Age world ensued, especially in northern latitudes. Both sea and land temperatures remained low between about 19,000 and 16,000 years ago. Thereafter, warming accelerated. But between 15,000 and 13,000 years ago, temperatures rose rapidly, perhaps as much as 7°C each century. The climatic seesaw then unexpectedly plunged again, with much colder temperatures between about 13,000 and 11,600 years ago. This cold "snap," the Younger Dryas (see Prolegomenon) lasted for just over 1,000 years. Temperatures plummeted, arctic vegetation returned to Europe, and ice sheets again advanced. Europe and Southwest Asia became much drier. A severe drought descended over much of the Middle East, forcing many bands to migrate in search of food. The causes are much debated, varying from a wave of volcanic activity to a potential meteorite impact. The effects of the Younger Dryas were regional and coincided in part with the first appearance of agriculture in the Middle East. The period ended with a return to a gradually warming earth, which has continued until today.

CHANGING LANDSCAPES (AFTER 16,000 YEARS AGO)

But what did this mean for our ancestors? In the north of Eurasia, after about 16,000 years ago, hunting bands expanded further north into newly deglaciated areas that had transformed into open steppe. As temperatures warmed after the Younger Dryas, birch, then ultimately oak, forests replaced these steppes.[3] European hunters turned from reindeer and cold-loving game to prey such as red deer, boar, and other forest animals. The hunter still used the spear and the spear thrower, a hooked stick that propelled missiles with unerring accuracy. Simple bows and arrows had come into use long before, maybe first in

Africa over 50,000 years ago, but they came into their own when new stonework techniques allowed the manufacture of small, razor-sharp arrowheads. These lightweight weapons, with their greater range, had a huge advantage: the ability to kill birds on the wing.

With the bow and arrow, rabbits, rodents, and migrating waterfowl became valued foods, captured not only with nets and traps but also with the new lightweight weaponry. Wooden arrows were tipped with small, deadly, sharp barbs and lethal heads that weighed almost nothing. These are known to archeologists as "microliths," or small stones. As the range of quarry widened, so did the exploitation of plant foods of all kinds. Cereal grasses, fruits, and nuts, now far more than supplemental foods, formed a central part of post–Ice Age diets. Many bands settled by lake shores and rivers and by sheltered ocean bays, where fishing and mollusk foraging assumed ever-greater importance. In many locations, groups large and small may have used the same camps year after year, and even permanently, depending on the seasonal abundance of foods.

The warming millennia brought dramatic changes to coastlines and rivers as melting ice raised global sea levels. Continental shelves, like that off Southeast Asia, vanished. The Bering Land Bridge between Siberia and Alaska became a stormy ocean. The North Sea between the British Isles and the Continent was once a land bridge formed of low-lying wetlands and lakes until about 8,500 years ago. Geologists call this sunken ancient world Doggerland, after the Dogger Bank, which is today a rich fishery.[4] Several thousand people once flourished there. Many bands must have spent virtually all their lives in wooden dugout canoes, casting fishing nets, laying fish traps, and hunting wildfowl, as well as deer and other small game. Like all Ice Age people, they were constantly on the move, but the imperatives of mobility were determined not simply by the movements of animals or the seasons of plant foods but by the changing water levels. In this near-flat landscape, a sea level rise—or even, as happened once, a tsunami—meant horizontal flooding. A sheltered canoe landing could vanish underwater in less than a lifetime.

The impact was profound. Animals chose new migration routes and moved on when their habitats became waterlogged. Sudden floods brought disease and new parasites. Above all, the loss of hunting grounds and well-defined band territories at a time of rising population densities caused profound social unrest, competition for newly scarce foods, and, inevitably, violence and warfare. The constant, seemingly unstoppable changes and environmental threats evoked a permanent sense of insecurity, even fear, as the threat of rising sea levels does on Pacific Islands and other low-lying parts of today's world. Every major flood within Doggerland meant the loss of a landscape previously imbued with meaning and emotional memories, with family histories and kin ties. It also meant a loss of practical knowledge: where to find the best fish or good flint. While some analysts may casually remark that mobility was a viable adaptation to climate change, there must have been times when the ecological changes engendered trauma, even crisis. Around 6500 to 6200 BCE, the rising Atlantic created the North Sea, which drowned the former Doggerland land bridge, with deep waters now separating Britain from the Continent.

Yet, as the world warmed, opportunism came into play, as it always has for humans. Neither tied down to permanent houses nor invested in growing crops, early humans would have found it relatively easy to keep moving, at least compared to later generations of settled folk. Moreover, people's intimate knowledge of the environment meant they could respond in resilient and creative ways to the changing climate. Certainly, we see hints of this in technological innovations, such as new types of fishing gear—for instance, a finely carved, many-toothed, bone harpoon point dredged up near the Dogger Bank in 1931.

The first people who crossed the Bering Land Bridge from Siberia to Alaska, sometime before 15,000 years ago, made one of the most remarkable adaptions to life in new landscapes.[5] The first migrants were arctic hunters, who spread remarkably quickly southward into North America and beyond, probably along the Pacific Coast. Within a few thousand years, though still thin on the ground, people had adapted to an amazing range of environments, from arctic tundra to vast open plains, deserts, and tropical rainforests.

Initially, the human population of the Americas was infinitesimal, widely scattered, and organized in small bands. The first Americans were transitory folk, constantly on the move, with only occasional contact with others. Their toolkits were light and easily carried; many of their hunting weapons and other devices were made when needed and soon discarded. What they left behind is virtually invisible, usually mere scatters of stone tools and flakes and occasionally animal bones. As far as we can tell, they relied on sharp-edged knives and stone-tipped spears, which bear little resemblance to those in Siberia, a sign that their new landscape led to new adaptations. A scatter of tools and radiocarbon dating go back as early as 14,000 years ago, perhaps earlier.

Around 13,000 years ago, the widespread Clovis people, known for their distinctive stone projectile points with thinned bases, appeared in North America. They were skilled hunters of animals of all sizes and also foraged widely for plant foods. Like their predecessors, they were highly mobile, following bison and smaller game over long distances. Clovis groups also acquired fine-grained toolmaking stone from afar. For example, flint Clovis points made from Knife River Flint from quarries in North Dakota have been found near Saint Louis, Missouri, 1,770 kilometers away. Their mobile, versatile bands adapted to all manner of challenging environments, from the grasslands of the Great Plains to the desert landscapes of the west, with temperature extremes ranging from northern cold to desert heat.

The Clovis cultural tradition flourished for about five hundred years. Then it gave way to another hunting-and-foraging tradition known as the Folsom, a cultural label for hundreds of small hunting groups that flourished over a vast area from the frontiers of Alaska to the Gulf of Mexico. Many groups pursued bison, but Folsom bands adapted to a wide variety of environments, from the Rocky Mountains to prairie woodlands east of the Great Plains. As the centuries passed, their successors adapted to all manner of natural environments, among them the desert west, the eastern woodlands, and exceptionally rich river estuaries and lakeshores, where increasingly elaborate hunter-gatherer cultures flourished at the same locations for many generations, relying heavily on fish and plant foods as well as game. Here kin ties and reciprocal exchanges of food and other commodities led to greater residential stability and close ties to ancestral lands. Some of these societies were the precursors of later, much more complex hunting and farming societies.

All these societies handled profound climate change, especially increasing aridity and greater warmth, as always, with a mixture of cultural values, instinct, and deliberate strategies like broadening the diet and mobility. Rising population densities and regular contact with other groups made it easier to share food, cooperate, and, above all, provide knowledge of the intricate landscapes for which they shared a universal respect.

A PERFECT STORM

As Ice Age temperatures warmed in earlier millennia, forests expanded rapidly across Southwest Asia—though temperatures were still cooler than today and rainfall was relatively plentiful. The vegetation became more diverse and included wild cereals that provided rich harvests of edible seeds. Game abounded; so did cereal grasses and edible nuts like pistachios and acorns. Especially along the lowest reaches of the Tigris and Euphrates Rivers, generations of hunters and foragers lived so well that they began to stay put. They founded increasingly larger settlements and buried their dead in cemeteries, interring many of them with lavish ornaments. There are signs of more elaborate social organization and, above all, of a more profound reverence for ancestors, those who had occupied the same land in earlier generations. This is not surprising: a good way to legitimize your claim to land is to emphasize your close ties to the ancestors whose land it had once been.

But why settle down in the first place? After all, for over 6 million years hominins—and for 300,000 years *Homo sapiens*—had been on the move. One scenario argues that, after the ice, the bountiful, clement conditions between about 14,500 and 12,900 years ago encouraged foragers to settle down permanently in villages near fruitful patches of land. Another posits that the population increase brought on by the increased rainfall and improved food availability meant people actively wanted to stake claims on "tribal territories." The reality was most likely a combination of both.

The warm period of plenty was, however, followed by the cold Younger Dryas, which brought not only drier and somewhat cooler conditions but also widespread drought in places like northern Lebanon. We've long known that the Younger Dryas affected foraging societies in Southwest Asia, but we now have the fine-grained details, thanks to research into speleothems at Soreq Cave in Israel and from other climatic proxies, including pollen and isotope records.

For humans, the drier conditions led to a greater emphasis on harvesting wild cereals and building storage facilities for them. At the same time, experiments with plant cultivation were well under way, having begun with barley and wheat at least temporarily as early as 23,000 years ago at the Ohalo II camp on the shores of Lake Kinneret in Israel. The experiment appears to have been short-lived and was abandoned when rainfall increased. Planting of wild grasses had apparently become a well-established strategy in arid environments where both plants and animals were unpredictable resources. No question, other groups cultivated cereals during the late Ice Age, but a widespread shift to deliberate cultivation came about as genetic changes in now-domesticated cereal grasses led both to full-time farming and to significant human population growth.

Yet the transition to food production was a far more complex adaptive process than it appears at first glance. No single individual "invented" agriculture or decided one day to domesticate useful animals. Instead, a gradual changeover unfolded independently in as many as fourteen places, probably more, often in response to the shift in climate.[6]

THE FIRST FARMERS (C. 11,000 YEARS AGO)

Various early experiments notwithstanding, food production began in earnest in Southwest Asia, East Asia, and South America about 11,000 years ago. Some 3,000 to 4,000 years later, there were farmers along the Yangtze and Huang He Rivers in China. Farming and animal husbandry took hold in South and Southeast Asia, in parts of the African savanna, and in North America 5,000 years ago. The new economies spread inexorably but at varying rates, depending on local environments. A more reliable source of food yielded a persistent growth in human populations and population densities. Only about five million people lived on earth when food production began, but this number boomed to between two and three hundred million people by the time of Christ. Today, subsistence and industrial agriculture supports 7.5 billion people worldwide, a number that's only growing. Yet fewer than one million humans still live by the age-old way of hunting and foraging.

Over half a century ago, archeologist Vere Gordon Childe wrote of two major revolutions in human history: The Agricultural Revolution and the Urban Revolution.[7] Childe's revolutions mask much more complex changes in human society generated by the ability to produce food. These included not only the development of expertise with crops and animals but also the establishment of larger, permanent settlements with much denser populations.

As a Marxist, Childe was particularly conscious of the social and economic issues associated with the settled life—our accumulation of property, investment in finite land, and subsequent domination of the few over the many. With settled life, there was indeed a strong trend toward competition, social inequality, and increasingly hierarchical societies. On the other hand, the new economy also meant that some were now freed from the everyday tasks of food procurement and could specialize in other things, like pottery making or metallurgy, or simply spend time thinking and focusing on other aspects of their lives. This is precisely why settled societies facilitated so many innovations in metallurgy, writing, art, and science. Moreover, as people multiplied, so did their ideas—especially when they coalesced in towns and cities and could share knowledge and thoughts. Populations rose not simply because of an abundant food supply (abundance was never guaranteed) but because multiple children—as the future labor force—were always a benefit in farming societies. Contrast this with hunter-gatherer societies, in which too many offspring would burden the group's food supply. As populations grew, villages became towns, towns became cities, and cities became kingdoms, then powerful empires.

This led to other unanticipated consequences: the appearance of new infectious diseases, incubated by domestic animals or insects, and stresses on the environment. These "changes of civilization" have had a fundamental effect on global climate. Looking back over the past 750,000 years, at least eight warm interglacials alternated with cold glacial periods, each of which began with high levels of greenhouse gasses that slowly declined as the temperature dropped. Then

Slash-and-burn agriculture in Tanzania, East Africa. Deforestation as a result of subsistence agriculture and animal husbandry has dramatically transformed global environments and carbon levels in the atmosphere. *Ulrich Doering/Alamy Stock Photo.*

came the current era, known to geologists as the Holocene, which is, of course, the era of farming. Climatologist William Ruddiman has shown how carbon dioxide levels fell gradually at first but then started to rise about 7,000 years ago.[8] Methane started to rise about 2,000 years later. He has argued that the increase of carbon dioxide was due to deforestation for agriculture and the rise in methane a result of wet rice farming. Ruddiman's theory is controversial but increasingly widely accepted. One can argue that this ancient shift from hunting and foraging to farming slowly, inexorably, and indeed, inadvertently fostered global warming and increased our vulnerability to short- and long-term climate change dramatically.

Vulnerability has always been there, of course. Short-term events like catastrophic droughts could descend without warming. How could the societies of the past adapt to and survive abrupt climate changes? Obviously, the quest for food was the overwhelming factor that drove society. When conditions were favorable and game and plant foods plentiful, decisions about survival were relatively straightforward and based on the most easily obtained sustenance, as well as influenced by competition with neighbors. When conditions worsened, new issues came to the fore, among them the minimizing of risk. Intuition played an important role, as did traditional survival strategies. Some might move to new places without strife; others might compete for resources, resorting to violence, with no guaranteed outcome.

A great deal must have depended on long-term social memories, knowledge about the environment and food resources, passed down from one generation to the next. Once tied closely to the land, farmers as opposed to hunter-gatherers were risk averse, knowing full well that repeated crop failures or animal diseases could make it impossible to adapt to, say, a prolonged drought. Many people must have died and groups passed into extinction as a result. Herein emerged an important distinction between foragers, who were constantly on the move, and those experimenting with cultivation, who would stay in one location for generations. Even the earliest farmers had a strong investment in their land and in their houses, storage facilities, and ritual centers. With such psychological attachment to the environment in play, they tended to react actively to environmental change, for example, by herding sheep rather than cattle. Abandonment of a settlement and its cherished lands was a strategy of last resort.

We can see the role of the environment in human life when we ponder the fact that farming began in earnest in the northern Levant during the rapid chill of the Younger Dryas.[9] This change in climate may have led to grain production because winter frosts killed seeds and delayed the germination and maturation of cereal crops. Groups were forced to alter their food sources. These were times of changing and unstable seasonal conditions. There were serious demographic pressures in landscapes that could only support small numbers of people. Intense social upheaval, competition for food, and countless small-scale migrations resulted. Foragers reacted by moving from more arid areas, like the Negev Desert and the margins of the Syro-Arabian Desert, to arable land. Within a short time, foragers could only survive in marginal areas close to the desert where cultivation was impossible.

The great changeover occurred in areas with Mediterranean vegetation or close to the steppe within the Fertile Crescent.[10] In other more forested landscapes, foragers continued to flourish alongside

Locations in the Middle East and Egypt mentioned in the text.

farmers. Between 11,700 and 11,200 years ago, farmers developed new forms of axes and adzes and used more efficient grindstones, stone-bladed sickles, and new, more effective arrowheads. Their settlements became more permanent and boasted adobe or brick walls with flat roofs, often on stone foundations. The earliest evidence of ceremonial buildings, like the shrines at Göbekli Tepe in southeast Turkey, dates to this period. That site's inhabitants, who turned an entire hilltop into a ceremonial center, were, as far as we know, still hunter-gatherers, not farmers. But they built an elaborate circular structure with carved stone uprights bearing animal figures that suggest it was an important sacred place.

The art, figurines, and plastered human skulls associated with such shrines reflect a strong obsession with ancestors as guardians of the land and with the powerful mythic beings that created the environment and the climatic forces that nourished it. These preoccupations reflect a greater concern with territorial control. At the same time, carefully fashioned animal or human figurines or wall decorations in shrines testify to frequent interactions with neighbors near and far. With these interactions came the sharing of cultivation and herding knowledge that allowed others to adopt new ways of subsistence and sustainability.

FIRST TOWNS: DRUGS, DROUGHT, DISEASE (C. 7500 BCE)

As forests retreated in the face of drought, wild grass harvests plummeted. Hungry communities survived off gazelle hunting and more intensive processing of grains and legumes. In areas like southeastern Turkey and northern Syria, some communities started planting wild grasses in attempts to expand their range, a familiar strategy of experimentation.

At a village mound named Abu Hureyra close to the Euphrates River in northern Syria, the original inhabitants of about 13,000 years ago lived in simple pit dwellings in a well-wooded environment, where animals and wild cereals were plentiful.[11] They also harvested hundreds of Persian gazelles, which migrated from the south each spring. By sifting the ashy occupation levels through fine screens, excavator Andrew Moore recovered large samples of plant foods. His colleague Gordon Hillman found that they came from half a dozen staple wild plants. But hundreds of others were also used for all kinds of purposes, including as hallucinogens and dyes. The tiny village was abandoned as the drought intensified, perhaps in part because of a shortage of firewood.

In about 9000 BCE, a new village rose on the low mound and grew to cover nearly twelve hectares. Within a generation or so, gazelle hunting gave way to sheep and goat herding. Hillman found that at first the people had foraged for fruit and grasses in the nearby forests. As the drought intensified, the wild grasses that once grew close to the houses became much rarer. Four hundred years later, the drought situation was much worse. At first the people adapted to what had always been a semiarid region by turning to small-seeded grasses and other standby foods. Judging from their skeletons, life for the first farmers was uniquely hard compared with before. Some young men had neck and spinal issues consistent with carrying overly heavy loads, like bundles of grain or building materials. Women typically displayed worn toe bones; a pathology consistent with a constant toe-curling/rocking position—the very pose needed to endlessly process grain on the grinding stones set into the floors of houses. Despite the issues, the population swelled, reaching as many as four hundred inhabitants. Living as they did in a now-denuded environment of dry steppe, they adopted humanity's long-established prefarming strategy: they eventually moved, abandoning their village, and fell back on better-watered landscapes.

As more favorable conditions returned after 7700 BCE, a much larger village of one-story mudbrick houses separated by narrow lanes rose on the mound. Abu Hureyra was not unique. As wetter conditions returned and the drier centuries were forgotten, farming and herding spread from coast to interior, from lowlands to uplands, through Mesopotamia into Turkey and the Nile valley. Yet climate change alone did not cause people to become farmers. The process was far more complicated.

Another long-term excavation has taken place in central Turkey at Çatalhöyük, a large village or small town of tightly packed houses, rebuilt at least eighteen times over 1,700 years, between about 7400 and 5700 BCE.[12] Daily life revolved around clusters of dwellings, occupied by the same families for many generations. Many are decorated in a lavish art style that displays elaborate symbolism. There are wall paintings of humans and dangerous animals and plastered skulls of humans and bulls. In the lived-in houses, the occupants interacted intimately with the past. Other houses contained the skeletons of far more people than had lived there. It is as if they were what the excavators called "history houses," places where rituals were performed and the living had access to the revered ancestors.

Not that life at Çatalhöyük was necessarily pleasant. Between 3,000 and 8,000 people dwelt in or close to the settlement at the height of its prosperity, when rainfall was relatively abundant and trade flourished. They experienced overcrowding, infectious diseases, violence, and serious environmental problems. What began as a small village in about 7400 BCE mushroomed rapidly into a densely populated, much larger village, even a town, which prospered off a lively trade in obsidian—volcanic glass prized for toolmaking. Today, bioarcheologists can study the chemical signature in the inhabitants' bones. The stable carbon isotopes show that the people relied on a diet heavy in cereals like barley, rye, and wheat. They kept sheep from the beginning, cattle in later times. Their grain-heavy diet led to numerous cases of tooth decay. The cross sections of peoples' leg bones reveal that the later occupants walked much more than the original inhabitants. The researchers believe this was because they had to farm and graze their flocks and herds much further from the community. Clark Spencer Larsen, who led the research, believes that environmental degradation and climate change forced community members to move away from the settlement both to grow crops and to gather adequate supplies of a vital commodity: firewood.

Çatalhöyük flourished during a period when the climate throughout the Middle East was becoming progressively drier. Inevitably, chronic crowding and poor hygiene led to infections, which appear in the bones of the dead. Dwellings were like crowded tenement buildings, so much so that analyses of the walls and floors yield traces of both animal and human feces. Trash pits and animal pens lay in close proximity to some houses. Sanitary conditions must have deteriorated rapidly. The crowding also led to violence. More than a fourth of a sample of twenty-five people had healed fractures. Some of them had suffered repeated injuries, many caused by blows to the head with hard clay balls hurled at them as they faced away from the assailant. More than half the victims were women. Most assaults took place during the most crowded generations of occupation, perhaps a time of stress and conflict within the community. Most striking of all, the kinds of problems faced by the Çatalhöyük farmers are virtually identical to those that are commonplace on a much larger scale in urban situations today.

Much more frequent and wider interconnections linked communities near and far at a time when people became ever more closely tied to their lands. A compelling philosophy of continuity became a central part of life in societies where routines now revolved even more intensively than ever before around the endless passage of the seasons. This was the context within which farming village societies almost everywhere were to handle the challenges of climate change.

Between 6200 and 5800 BCE, catastrophic droughts affected farming communities between the Euxine Lake (now the Black Sea) and the Euphrates River. The aridity was unrelenting; lakes and rivers dried up; the Dead Sea sank to record low levels. Farming communities large and small shrank and withered in the face of pitiless drought. Many vanished, and numerous people died of starvation and famine-related disease. Others, like the inhabitants of once prosperous Çatalhöyük, turned from cattle herding to sheep farming, unable to satisfy the water needs of herds of the larger animals.

LIVING PRECARIOUSLY

The end of the Ice Age required humans to adapt to climatic shifts in dramatic ways. Yet, as the ice sheets retreated, the seas rose, and the landscapes altered unimaginably, humans (who were still nomadic hunter-gatherers) drew on what they knew best: they adapted in the traditional way, relying on experience, kin ties, and cooperation, as well as technological ingenuity, to reduce risk and maintain resilience in the face of profound cultural and environmental change.

The greatest changes came with the move to agriculture and animal husbandry in different parts of the world after about 11,000 years ago. Subsistence agriculture and herding tied people to their lands, and they began to "import" goods unavailable locally. At this point long-distance trade, with its "exotic goods," really took off. Obsidian, a fine-grained volcanic glass, became highly desirable for toolmaking and for ornaments. British archeologist Colin Renfrew used distinctive trace elements in the rock to trace obsidian trade routes over a wide area of the eastern Mediterranean.

Yet most farming-dependent people had it tough and lived precariously. For the first time, humans lived with the harsh realities of subsistence agriculture, of short- and long-term drought in landscapes from which they could not move, as they used to. It requires far less time and energy to simply go out and gather food or hunt beasts than to invest in the land and farm. Anthropologists have demonstrated this over and again in their work with hunter-gatherer groups, such as the Hadza foragers of Tanzania, closely documenting the calorie expenditure and calorie return of their lifestyle. Moreover, as anthropologists have demonstrated with the San of Africa's Kalahari Desert, farming requires far more calories and time than is ever needed to collect the equivalent amount of food among hunter-gatherers—not to mention the fact that hunter-gatherers keep their population numbers down, thus requiring less effort from the group.

But those farming groups who prevailed did so thanks in considerable part to the legacies of risk management that had been passed down through generations. Adaptation to climatic variation was a local concern, which depended on environmental knowledge and inherited experience. As still-settled people, we often forget the importance of local adaptation. Measures against climate change always started at the local level and were appropriate to the surrounding local environment. This remains true today, whether we live in a village or in a city with a population of millions.

And denser populations were already on the horizon as farming economies expanded. If "success" is based on increasing population densities, then farming was brilliantly successful. Within a few centuries, farming communities dotted the landscapes of northern and southern Mesopotamia, "the Land between the Rivers." Soon towns, then sophisticated cities with writing, monumental architecture, gold, jewels, charismatic kings, and full-scale warfare would emerge. The next two chapters explore how Mesopotamia and its contemporary civilizations in Egypt and the Indus valley worked, and failed, as they battled the sun and rain.

3

MEGADROUGHT

(c. 5500 BCE to 651 CE)

Marduk, king of the gods and of humankind, lord of justice, health, farming, and thunderstorms, presided over the primordial cosmos between the Tigris and Euphrates, the great rivers of Mesopotamia. Or so ancient legend tells us. He mounted his storm chariot and forged order out of chaos with floods, lightning, and tempests. The charismatic deity overcame the dragons of chaos and fashioned the tumultuous spiritual and human world of the Sumerians, the first city dwellers on earth. Marduk presided over a land of climatic extremes, with baking summer temperatures as high as 49°C and winters with violent, heavy rains and chilling temperatures. His world was tumultuous, volatile, and always unpredictable.

His fellow gods created the cities that competed for power in this fertile and violent land. "All lands were the sea, then Eridu was made," proclaims a Mesopotamian creation legend, inscribed on a clay tablet many centuries later. The city of Eridu, west of the Euphrates River, lying in today's Iraq, was indeed the oldest town of all, the dwelling place of the god Enki, lord of the abyss and god of wisdom. Eridu's earliest shrine dates to about 5500 BCE, found under a magnificent, stepped ziggurat (temple mound) adorned with brightly colored bricks five centuries later. The rapid growth of another city, Uruk, again in today's Iraq and also close to the Euphrates, accelerated after 5000 BCE, when two large farming villages coalesced into a single settlement.[1] Uruk was the home of the mythic hero Gilgamesh. Nearly 2,000 years later, by 3500 BCE, it was far more than a large town. Satellite villages, each with its own irrigation system, extended about ten kilometers in all directions. Four centuries later, Uruk covered nearly two hundred hectares, a city of between 50,000 and 80,000 people. It became a major religious and trading center connected to a far wider world flanked by the major trade routes of the Euphrates and Tigris Rivers. An imposing shrine, said to have been dedicated by Gilgamesh himself to Inanna, the deity of love, lay at the spot where the goddess herself was said to have planted a willow tree taken from the waters of the Euphrates. According to *The Epic of Gilgamesh*, Uruk had four quarters: the city itself, gardens, brick pits, and the largest, the temple quarter.

The goddess Ishtar's ziggurat and temple precinct lay amid a crowded metropolis of teeming neighborhoods of mudbrick houses. Most were close-knit communities of fellow kin with long-standing ties to villages in the hinterland or precincts where expert artisans lived and worked. Narrow streets separated dwellings, wide enough for loaded beasts. On calm, cold days, a pall of smoke from domestic hearths and workshops mantled the city with its busy markets. Uruk was a cacophony of animal and human voices—dogs barking, vendors touting their wares from their stalls, men quarreling and women walking together to buy grain, distant chants from the temple enclosure. It was a mélange of smells—foods, cattle manure, decaying waste and urine—but like all Mesopotamian cities, it was a vibrant place in a landscape of potentially dangerous natural events.

Cities soon became the norm.[2] By the end of the fourth millennium BCE, more than 80 percent of the southern Mesopotamian population lived in settlements that sprawled over ten hectares in a turbulent land ruled by intensely competitive city-states. They formed what we call the Sumerian civilization, centered between the Euphrates and the Tigris in what is now southern Iraq. Sumer was very rarely a unified state, in practice little more than a patchwork quilt of cities and city-states dependent on Indian Ocean summer monsoon rainfall, in addition to spring and early-summer river floods.

As cities grew, agricultural production soared to feed thousands of nonfarmers. Such farming depended on spring and summer floods down the great rivers. After about 3000 BCE, the Indian Ocean monsoon that brought summer rainfall weakened. The rains faltered, began later, and ended much earlier. Rainfall also declined in Turkey, the source of Euphrates and Tigris floods. The Mesopotamian climate became less stable, with prolonged drought cycles that played havoc, especially with smaller communities dependent on rivers that continually and unexpectedly changed course.

Even with plentiful rainfall, this was a challenging setting for irrigation farming.[3] With the growth of cities, the demand for grain and other staples skyrocketed. For centuries, farmers had cultivated narrow bands of fields along the backslopes of natural levees and along the edges of flooded depressions. They also took advantage of natural levee breaks and the better drained sediment deposits they spawned, which allowed small-scale irrigation. But such fields could only support relatively small settlements, mainly villages extending along main channels that were also trade routes. This was one major reason why agriculture was most effectively managed at the local level.

In the face of drier conditions, intensive agriculture and human-made irrigation works were inevitable as cities soon housed between 5,000 and 50,000 people. The already close webs of interdependency that linked village to village and villages to the city assumed even greater importance. Climate change and growing numbers of non-farmers meant that agriculture based on square, level plots gave way to blocks of more standardized long fields that required close supervision but were plowed by oxen to create furrows. A farmer's almanac of the third millennium gave explicit watering instructions: "When the barley has filled the narrow bottom of the furrow, water the top seed."[4] There was no central authority, as was the case much later when industrial agriculture developed in the West during the nineteenth century CE. Instead, agriculturalists comprised a patchwork of small tribes, each with its own irrigation works of varying size, which shifted constantly as they adapted to rapidly changing conditions. To manage this economic palimpsest under local control required an intimate knowledge of village politics and rivalries, a major challenge for any centralized authority.

At first there were no despotic kings, no powerful rulers to dictate policy or to allocate water or repair channels. Power lay in the hands of tribal chiefs, whose authority depended on the loyalty of villagers, on kin relations, and on the kinds of reciprocal ties that linked every member of rural and often urban society. These social and political realities made for chronic tensions between cities and their hinterland communities and for endemic turbulence and unrest, which endured long after Sumerian civilization ended.

As urban populations swelled, the pressure for ever-larger grain surpluses intensified. Strip fields and their closely packed furrows required a level of organization that transcended families and kin groups. A new element came into play—a form of social authority, perhaps based on temples, which supervised irrigation and farming on a wider scale. It's easy to imagine taxation, when what actually developed was a corvée assessment that provided labor not only for irrigation works but for public works of all kinds. Laborers were paid in carefully allocated rations, which in turn pressured farmers to meet tax assessments. As proof of such labor, standardized ration bowls with beveled rims for workers appear from northern Mesopotamia deep into Iran. While temples, like that of Ishtar in Uruk, became powerful economic, political, and social forces, Sumerians lived in a bipartite world of cities and villages. The villages produced food; cities were centers of manufacturing, trade, and religious activity. A third millennium proverb aptly remarked, "The outer villages maintain the central city." Another tablet commented, "The man to fear is the tax collector."

Italian scholar Mario Liverani has written of the major step that transformed Mesopotamian farming communities.[5] For centuries, they had lived at the self-sufficient, subsistence level. Soon they became what he calls the "outer circle" of newly formed urban societies. Farming communities supplied labor for food production and urban development projects, without receiving much, if anything, in return except the satisfaction of serving whichever patron god presided over the nearby city. The food and labor generated by the outer circle fed an inner one of artisans, officials, and priests, who received rations. This unequal way of food production and redistribution inevitably created cities based on social inequality and privilege. The inner and outer delineation soon prompted a divide between elites and commoners and a system reinforced by ritual and "wisdom literature" that emphasized cooperation through hierarchy. One proverb urged, "Don't drive out the powerful, don't destroy the wall of defense."[6]

SUMERIANS AND AKKADIANS (C. 3000 TO C. 2200 BCE)

Sumerian ideology spoke of two great irrigation channels, the Euphrates and the Tigris, flowing from the northern mountains to the urban south of Mesopotamia. It also told of gods and rulers who carried hoes and baskets, as if they themselves tilled the soil, giving religious value to the provision of food and agriculture. Everything in the south depended on irrigation, which meant that every farmer knew the subtle features of the floodplain—the locations of the most fertile soils and places where flood waters would habitually breach natural levees. Judging from later inscriptions, the best farmers learned the telltale signs of potentially catastrophic floods and of impending low-water years.

Mesopotamian farming had never been easy, even during centuries of better rainfall. Initially at least, permanent irrigation channels were an impossibility, for stream channels shifted constantly and without notice. River course changes were a constant hazard, but unexpected natural levee breaks also provided opportunities to direct water onto potentially fertile land.

Farming became even harder as conditions became drier after 3000 BCE and urban populations increased. The informal, volatile village irrigation systems of the past gave way to more formal approaches, which were, however, still community based. They could be no other way, given the dependence of cities on village food surpluses. Sumerian society came under the rule of secular rulers, *ens* or *lugals*, who presided over agriculture, war, trade, and diplomacy.[7] The ever-changing maze of political alliances and individual or kin obligations that had linked communities for centuries now operated on a much larger scale as political power passed into fewer hands. The changing river systems and the concentration of settlements in major irrigation areas heightened diplomatic and political questions in a world where a ruler in a strategic location could potentially deprive his neighbors of water and starve them to death. Cities like Lagash, Umma, Ur, and Uruk quarreled violently over water and agricultural land. The rhetoric was as shrill in 2500 BCE as it is today: "Be it known that your city will be completely destroyed! Surrender!"[8] Fragmentary records of disputes over control of water supplies and farmland speak frequently of "raising the battle net of Enlil," for the battles were invariably fought in the names of the gods. The great rivers meandered over the landscape in broad loops and sometimes changed course when they burst their banks, a crucible of intercity strife and war. By 2700 BCE, walls appeared around many cities like Lagash and Ur (the biblical Ur of the Chaldees). Cycles of boom and bust, population rises and declines, and rises in soil salinity caused in part by shorter fallow cycles reduced crop yields—at Ur by half from earlier times.

Increasing aridity and a need for larger food surpluses made year-round farming an essential routine. Cities like Ur and Uruk forged organized networks of trading contacts that extended up the great rivers far into Turkey and had such political and cultural impact that they formed what Mesopotamian specialist Guillermo Algaze has called the "Uruk World System." Sumerian lords competed with cities as far northwest as Syria. They attacked trade routes and annexed neighbors, but such moves were fleeting, for internecine strife and petty rivalries at home intervened. Inevitably, some rulers developed broader territorial ambitions. In 2334 BCE, Akkadian king Sargon of Agade, south of Babylon, defeated a coalition of Sumerian city-states led by King Lugalzagesi of Ur.[9] He crafted the first-known empire, which encompassed all of Mesopotamia and lands far to the west, east, and south. But his enlarged and loosely controlled domains were far more vulnerable to severe drought than their smaller, volatile predecessors. Ultimately, agricultural production depended almost entirely on local officials and community leaders.

Sargon and his successors created an empire that depended on the loyalty of their officials, the liberal use of patronage, and the toil of thousands of commoners and war captives, for, like all preindustrial civilizations, the Akkadians relied on raw human labor. The increasingly elaborate superstructure of the empire required carefully apportioned rations not only for unskilled laborers but also for high officials, skilled artisans working in cities and palaces, and entire armies of conquest. Almost all Akkadian military campaigns and subsequent exploitation of new lands depended on conquered cities and villages in the north and south that provided huge food surpluses. The Akkadian rulers' power depended on this network, while two elements of the ecosystem were particularly important: good rainfall in the north and river floods to nourish the fertile soils of the south.

We know from scanty cuneiform records that Akkadian officials carefully monitored flood levels, concerned as they were about crop yields and rations. However, there are no signs of any long-term concern about vulnerability to prolonged drought. Akkadian imperial activity reached its height in about 2230 BCE but lasted less than a century, when, without warning, the rains failed. Rainfall dropped to between 30 and 50 percent of normal. A megadrought ensued. It lasted for three hundred years.[10]

A TERRIBLE DROUGHT (C. 2200 TO 1900 BCE)

The Great Megadrought of c. 2200 to 1900 BCE, often called the 4.2 ka Event, was a global climatic occurrence. This unprecedented arid cycle affected human societies from the Americas to Asia, from the Middle East to tropical Africa and Europe.[11]

Why did the megadrought occur?[12] We cannot be certain. Variations in solar irradiance and cycles of volcanic forcing have accounted for much of temperature variability over the past millennium. This may also have been the case in earlier times, but the North Atlantic Oscillation (NAO) was now (and still is) a major climatic driver. This giant atmospheric seesaw between the subtropical high over the Azores and the subpolar low accounts for up to 60 percent of the December–March temperature and rainfall variability across Europe and in the Mediterranean region. Since the NAO regulates Atlantic heat and moisture funneling into the Mediterranean, both Atlantic and Mediterranean Sea surface temperatures influenced, and still influence, Middle Eastern climate. There seems little question that the NAO seesaw contributed to the great drought.

The megadrought can be observed in lake sediments from Iceland and Greenland and in European tree rings. High-resolution speleothem sequences taken from caves as far afield as Turkey and Iran also recorded the event. Likewise, the three-hundred-year episode appears in East African and Indus valley paleoclimatic sequences, when the Indian monsoon faltered. There were sudden changes in Nile River floods and in rainfall along the Indus, as well as in the Sahara and West Africa. The unstable East Asian monsoon also created stress for well-established farming communities in eastern China.

The effects of the megadrought rippled across kingdoms, prosperous civilizations, and rural landscapes. As we shall see in Chapter 4, the megadrought coincided with the end of the Egyptian Old Kingdom and the temporary fragmentation of the pharaohs' domains. The effects of the drought extended as far as Tibet and into the Americas, where arid conditions coincided with the introduction of maize agriculture into the Southwest and the Central American Yucatán. The drought was also a factor in the rise and fall of important communities in the Andean region of South America.

In the Middle East, Dead Sea lake levels are thought to have dropped by about forty-five meters. The drought can also be identified in a marine core from the Gulf of Oman, while a speleothem sequence from Mawmluh Cave in northeastern India links reductions in Nile flow with falls in East African lake levels and the deflection of the Indian monsoon. As one would expect, the drought's effects varied considerably from one area to the next. In western Asia and northern Mesopotamia, vital dry farming areas contracted abruptly between 30 and 50 percent. Much of the Khabur Plains of northeastern Syria, the eastern Mediterranean, and northern Iraq suffered catastrophic drought.

The ways in which the unfortunate Mesopotamians coped with the drought also varied considerably. In dry farming areas like the Khabur Plains of the north, important centers like Tell Brak and Tell Leilan were abandoned completely.[13] These dispersals affected as many as 20,000 people in each city, and subsequently major building projects ceased. At Tell Leilan, Yale University archeologist Harvey Weiss excavated a large-scale grain storage and distribution center abandoned abruptly in about 2230 BCE. Partially constructed structures stood across the stone-paved street outside, signifying a desertion of city development. Here and elsewhere, clear administrative decisions were made to abandon important buildings. No one lived on the Khabur Plains after 2200 BCE until the rainfall situation improved 250 years later. Dry farmers abandoned major cities and other communities from the Upper Euphrates drainage in Turkey to the southern Levant.

Many dry farmers adapted to the drought by following (commonly known as "tracking") better-watered habitats southward to locations where springs nourished fields. But major coastal Mediterranean cities like Byblos and Ugarit, which lacked such water supplies, saw significant population reductions. Meanwhile, Jericho to the south benefitted from a natural spring for its large flocks of sheep. The Euphrates, while much reduced, still allowed some irrigation in central and southern Mesopotamia. Increasing aridity, however, also led to a boom in herding. Nomadic pastoralism became popular, a survival mechanism triggered by the interruption of ancient seasonal herding movements between the Khabur Plains and the Euphrates. The Khabur drought forced the nomads, known as Amorites, onto the nearby steppe, to the banks of the Euphrates, and southward into settled lands. Constant turmoil erupted as their flocks encroached on settled farming lands. So serious was the threat that the ruler of Ur built a 180-kilometer-long wall named "The Repeller of the Amorites" in about 2200 BCE to keep the visitors in check. His efforts were futile.[14] Ur's hinterland witnessed a threefold increase in population at a time when the city's officials were frantically straightening irrigation canals and issuing minute grain rations. Cuneiform tablets tell us that Ur's agricultural economy eventually collapsed.

In the south, the climatic shift was blamed on the gods as expressed in poems or "city laments." "The Lament for Sumer and Urim" is one of the earliest written records of the invocation of divine actions to explain climate change. We learn that Enlil, Enki, and other deities decided to destroy a city. "The storms gather to strike like a flood . . . [so] that the cattle should not stand in the pen, that the sheep should not multiply in the fold; that watercourses should carry brackish water."[15] They ordered "bad weeds" to grow on the banks of the Tigris and Euphrates and turned cities into "ruin-mounds." Crops could not be planted, and the countryside would dry out; "Enlil blockaded the water in the Tigris and Euphrates."

NEO-ASSYRIANS (883 TO 610 BCE)

Total destruction leveled entire Mesopotamian cities as crops died "on the stalk" and corpses floated in the Euphrates. Food was scarce;

canals silted up. Centuries of volatile political maneuvering and rivalry ensued until the ninth century, when Assyrian monarch Ashurnasirpal II (r. 883–859 BCE), the ruler of Mesopotamia's dominant Assyrian Empire, embarked on ruthless campaigns of expansion during a time of greater plenty. Any sign of revolt brought draconian punishment in an empire forged entirely by force. He appointed governors of unquestioning loyalty to control vanquished territories, demanding harsh tribute in precious metals, raw materials, and commodities like grain. After campaigning as far west as the Mediterranean, he returned home during a period of increased rainfall, identified in a speleothem in northern Iran, and used prisoners of war to build a magnificent palace at Kalhu (Nimrud) on the Euphrates. Then he threw a ten-day party around 879 BCE to celebrate its completion.

It was quite an event.[16] Ashurnasirpal boasted that 69,574 guests attended, including 16,000 from Kalhu itself. They feasted on thousands of sheep, oxen, and deer, birds, fish, all kinds of grains, and 10,000 jugs of beer and 10,000 full wine skins. The king sent them home replete, bathed and anointed with oil "in peace and joy." As Ashurnasirpal's visitors feasted, they gazed on walls adorned with brightly colored bas-reliefs inscribed with cuneiform. Twenty-two lines listed the king's credentials, while nine commemorated his victories. He was a "chosen one" of the gods Enlil and Ninurta, "great king, strong king, king of the universe . . . fearless in battle . . . trampler of all enemies." Unrelenting propaganda trumpeted his mastery of an Assyrian Empire created by murderous conquests—men, women, children all perished at his hands. Yet, a mere 270 years later, the wine-guzzling, ear-chopping Ashurnasirpal's empire imploded.

The Neo-Assyrian Empire (as archeologists call it) was the largest and most powerful empire of its time, in full throttle around 912 BCE, whence Ashurnasirpal's lavish celebration. Yet it grew even more powerful in the mid-700s BCE, under the formidable Tiglath Pileser III, the man who engineered Mesopotamia's largest expansion. His name turns up in everything from ancient Yemeni inscriptions to hostile remembrances in the Old Testament—notably for his invasion of Israel, capture of Galilee, and unfair tax extraction. Given such all-powerful kings, why did the Neo-Assyrian kingdom suddenly disintegrate in 610 BCE?

Did a series of bloody civil wars and insurrections destabilize the ruler's authority? Or did brutal warfare and military defeat undermine an overextended empire? Certainly, both played important parts. Assyrian rule, like that of earlier monarchies, was always fragile, being perennially volatile with none of the carefully nurtured precedents that supported the Egyptian pharaohs. However, we now know that another, familiar guest was at the table: climate change.

Kuna Ba cave in northern Iran tells the tale with a high-resolution, precisely dated speleothem record of climatic shifts.[17] The speleothems show that the Neo-Assyrian Empire's rise occurred during two centuries of unusually wet climate. The ample precipitation was a godsend to the thousands of farmers who provided food to cities and conquering armies dependent on the state's carefully apportioned rations. Thereafter, a series of megadroughts during the early to mid-seventh century, each lasting for decades, seems to have triggered a decline in Assyria's agrarian productivity, which in turn contributed to its eventual political and economic collapse. Ultimately, the entire Neo-Assyrian Empire finally crumbled amid bitter fighting—but with a people already weakened by drought.

TRANSFORMED LANDSCAPES

With cities and long-distance trade networks came increased demands for raw materials of all kinds, especially timber and metal ores. An insatiable demand for clay vessels, as well as for metal tools and ornaments, led to a constant demand for wood fuel for kilns, in addition to the wooden beams and other timbers used for construction of all kinds. Firewood was also always in demand, imported in large bundles by pack animals. The promiscuous use of wood for domestic and industrial purposes must have created dense clouds of wood smoke that would have hovered over the growing cities on calm days. Serious air pollution must have afflicted crowded cities, but the damage caused by deforestation had serious, long-term consequences.

The vegetation history of the Middle East is still little known, but the effects of near-industrial consumption of timber transformed most areas. In central Anatolia, for example, pollen diagrams show that the landscape was open oak woodland until between about 5000 and 3000 BCE, when tree coverage declined rapidly, as was the case in modern-day Iran and Syria. Kaman-Kalehöyük, one hundred kilometers southeast of Ankara, was a major settlement occupied during the second and first millennia, until about 300 BCE, and a major agricultural center with some textile and pottery manufacturing. The occupation coincided with a major drought between about 1250 and 1050 BCE, a period when the powerful Hittite Empire disintegrated. A study of wood charcoals shows that the Hittite inhabitants of the settlement harvested the surrounding woodland so intensively that the wood cutters did not use the well-established oak woodland of earlier times but instead exploited woodlands with a lower diversity of other species.[18]

MIGHTY WORK DISSOLVES (224 TO 651 CE)

After the megadrought, Mesopotamian civilization thrived anew, once rainfall returned to its previous seasonality. People repopulated the Khabur Plains and Assyria. Tell Leilan prospered once more. The diluted ideologies and institutions of earlier times survived to become the blueprint for the great kingdoms that rose on the foundations of earlier city-states. The new empires turned irrigation agriculture into a state business. Its roots still lay in the hands of local sheikhs and village farmers managing water and crops as they had done for many centuries, with all the volatility and constant feuding that had undermined Sumerian, Akkadian, and Assyrian rule. Those who cultivated Mesopotamian soil were fiercely independent and had no illusions about the natural landscape, which was by now drastically modified by human activity. They were well aware of the difficulties that confronted them locally quite apart from drought: chronic silting in irrigation canals and rising soil salinity among them. But agriculture was still stable enough to support the largest empire of the ancient world, that of the Achaemenid Persians (550–330 BCE), who existed in comparative peace and are known for their architectural gems, such as Persepolis.

Fast-forward to 224 CE, when the Sasanians created Persia's final pre-Islamic empire, which flourished for four centuries.[19] They controlled a vast area between the southern Caucasus Mountains and portions of the Arabian Peninsula. The central government used draconian policies that had served the Assyrians well, but on a much larger scale. The authorities invested heavily in irrigation systems that dwarfed earlier water-management efforts.[20] Like the Assyrians, they resettled deported populations in areas that appeared to have potential. There they founded new towns and embarked on massive hand-dug irrigation works to support them. One irrigation scheme, drawing on two rivers and built in the sixth century, carried water more than 230 kilometers to the Tigris. The scheme irrigated about 8,000 square kilometers of farmland northeast of Baghdad but brought water onto land with sluggish drainage. Long after the Sasanians, intensive land use resulted in acute salinization; many hectares languished. The scheme was abandoned by 1500.

The Sasanians brought an area of about 12,000 square kilometers between the Tigris and Euphrates under at least sporadic irrigation during the sixth century. This meant that they were cultivating an area at least double that of earlier times. Harnessing the Tigris was a high-risk venture, given the river's swift and variable current. The grids of canals and fields covered areas that were far too large for village farmers or a small city-state to control. The scale of the new water works meant that farmers living some distance from water sources were in big trouble if breaches occurred upstream. This was centrally planned, standardized irrigation on an unprecedented scale, driven by potential tax yields rather than harvests, aiming for maximal fiscal return in both grain and land taxes for the central government, not to satisfy local needs. Most of the canals were built by thousands of war captives, with the aim of resettling conquered peoples. The Sasanians moved away from smaller-scale irrigation works that took account of local conditions. They created predominantly artificial irrigation systems that yielded plenty of grain at first. But they ran into trouble as silt accumulated in poorly designed canals. Each elaborate irrigation scheme, each new demand from outside, reduced the self-sufficiency of rural commoners—those who worked in the fields. The aggressive Sasanian engineers focused on the short-term and neglected the all-important sluggish drainage that had preoccupied earlier farmers. At first, lavish returns brought prosperity and more revenue. But rising maintenance costs overwhelmed the engineers. Their newly constructed levees disrupted existing drainage patterns, raised groundwater levels, and created chronic salinity in agricultural land. Soon short-term increased yields came at the expense of increasing ecological fragility. Productivity dropped sharply, especially in marginal areas. Irrigation schemes lost their flexibility in the face of drought, high floods, and other climatic shifts. The bureaucracy surrounding agriculture and water management declined as economic and political weaknesses led to an impoverished agricultural population and centrally managed irrigation collapsed. The Sasanian Empire dissolved in the face of expanding Islam between 632 and 651 CE. By the eleventh century, the land between the rivers was an abandoned, salt-infested wilderness.

The Assyrians, Akkadians, and Sumerians lived through the dawn of an era where both rural and urban populations became ever more vulnerable to abrupt, often short-term climate change. Centralized government and authoritarian rule, like that of the Sasanians, did not solve the problem of growing population densities and erratic water supplies, whether from flood or rainfall. As was clear even as early as Sumerian times, the best solutions were at a local level, where individual community leaders could adopt smaller-scale measures to combat hunger. They knew the land, the vagaries of changing floods, and the mood and expertise of their people. When the complex relationship between city and countryside developed from interdependence to dominance, centuries of turbulent experience combined with independence made any long-term way of coping with severe drought or other climatic changes practically impossible. Doubtless some long-forgotten Mesopotamian leaders dealt with the challenge of intense drought within the narrow confines of their own domains, be they cities or provinces, but no records of their actions survive.

Mesopotamia lay between two great rivers, but the wider geography meant permeable borders and often unreliable land-based infrastructure. A persistent fluidity, loose control, and the ebb and flow of changing loyalties, combined with shifting patronage and royal ambitions, contrasted dramatically with the experience of the pharaohs along the Nile. One enduring lesson about adapting to climate change emerges: conquest and exploitation are not solutions, even if King Ashurnasirpal and Tiglath Pileser III thought they were. The historical experience in Mesopotamia resonates powerfully in today's world. Adaptation to changing circumstances, including climate change, is often most effective when the solutions are local rather than imposed by a remote bureaucracy or a large industrial enterprise.

4

NILE AND INDUS

(3100 to c. 1700 BCE)

The Greek historian Herodotus wrote of Egyptian farmers in the fifth century BCE, "They gather their crops with less effort than anyone else in the world . . . [T]he river rises of its own accord and irrigates their fields, and when the water has receded again, each of them sows seed in his own field and sends pigs into it to tread the seed down."[1] Every summer, heavy monsoon rains in the Ethiopian highlands swelled the waters of the Blue Nile and Atbara Rivers far upstream. The silt-heavy flood surged northward, reaching its height over about six weeks between July and September. Each year, *akhet*, the inundation, spread over the floodplain, which sloped gently away from the main river channel. This was the time of anticipation. A Pyramid Text remarks, "They tremble that behold the Nile in full flood. The fields laugh and the riverbanks are overflowed. The god's offering descends, the visage of the people is bright, and the heart of the god rejoices."[2]

An idyllic portrait from both Herodotus and the Egyptian scribes to be sure, but a misleading, indeed mythical, one. In reality, Egyptian villagers toiled endlessly to direct floodwaters to their fields with dikes and canals, which could melt away in the face of an aggressive flood. Ancient Egyptian farmers lived at the mercy of the Nile and of the distant interactions between the ocean and atmosphere that drove the Indian Ocean monsoon.

Nevertheless, they seemed to live in a timeless world, where the sun passed across the cloudless heavens day after day. Water, earth, and sun: these were the eternal verities of ancient Egyptian civilization.[3] The god Atum, "the completed one," was the creator. He emerged from the watery chaos of Nun, the primordial waters, and raised an earthen mound above the waters. But the sun god Ra was the supreme manifestation of power, appearing without fail at sunrise, moving through the heavens just as life moved forward. Egyptian belief and ideology depended on stable, wise government by pious pharaohs who presided over a harmonious state. Egypt's kings ruled as Horus, a manifestation of heavenly power and the skies, the epitome of good order. Their enemy was the snouted creature Seth, the essence of chaos and disorder. He brought storms, drought, and hostile foreigners to the harmonious Nile world. The conflict between Horus and Seth symbolized the forces of order and harmony, anarchy and chaos. Decisive, vigorous rulers with personal charisma symbolized a unified Upper and Lower Egypt—the Two Lands. The unification of the Egyptian state took centuries but was always (wrongly) presented as an act of harmony, a triumph of order over chaos.

Egypt was a linear civilization, hugging the fertile river floodplain and isolated from what was long perceived as a turbulent outside world. The pharaohs ruled according to precedent and were considered to be the personification of *ma'at*, which approximates to the modern word "order" or "rightness," qualities embodied in a wise and harmonious goddess of the same name who regulated both the seasons and the law. *Ma'at* stood in opposition to *isfet*, the forces of disorder. Egypt's demigod rulers governed by their own word, following no written laws and their version of holy writ. A massive hereditary bureaucracy—often veritable dynasties of major and minor officials—effectively ruled the state for them. Most of the time, the state worked. This was an extraordinary civilization that survived in various forms for over 3,000 years—supported both by *ma'at* and also by its unique Nile environment.

IN THE BEGINNING (C. 6000 TO 3100 BCE)

When farming began in southern Mesopotamia around 6000 BCE, and as Doggerland vanished under the North Sea, the Nile flowed through a lush river valley hemmed in by deserts. West of the Nile, irregular rainfall was sufficient to support rolling Saharan plains of arid grassland. Only a few thousand people lived in the valley: hunters, foragers, and fisherfolk, who may have cultivated some cereal crops. They traded sporadically with nomadic cattle herders from the desert, who came to trade or graze and water their beasts. The visitors' leaders were men of long experience and exceptional ritual ability, who were apparently expert rainmakers. Presumably, this gave them unusual credibility in an arid land.

The herders moved gradually eastward onto the Nile floodplain when the rains became more sporadic after 5000 BCE. As the Sahara grew ever drier, they settled permanently by the Nile, bringing with them new notions of leaders as strong bulls and herdsmen and perhaps rituals that led to the worship of the fertility goddess Hathor. Ancient Egyptian civilization had deep roots in the village cultures of earlier times, which depended on careful water management and the backbreaking labor of irrigation agriculture. A tradition of authoritarian leadership, perhaps inherited from village headmen, was already deeply engrained in the Egyptian psyche in a world where rainfall was virtually nonexistent. Everything depended on the life-giving flood and the confident leadership of a herdsman.

The Nile flowed through harsh deserts. From space, it looks like a green slash that heads like an arrow toward the Mediterranean to the north. The Egyptians called its floodplain *kmt*, or "the Black Land," its fertile black soils contrasting with the "Red Land" of the desert. Each year, the gods willing, the Nile brought water and silt far downstream, from its two tributaries, the White and Blue Niles, which flow out of East Africa and the Ethiopian highlands before uniting in Khartoum in today's Sudan to form the Nile itself. *Akhet*, the season of inundation, unfolded during the spring and summer months when the Nile flood spilled over the floodplain. The receding waters nourished the fertile soil for the farmers who planted their crops with carefully dug and maintained irrigation channels. Unlike in Mesopotamia, *Akhet* fertilized the floodplain and eliminated the risk of salinization. Farming was brutally hard work but relatively easy by Mesopotamian standards, without any need for fallowing or fertilizing the fields. The farmers merely steered the rising waters through canals and reservoirs built to retain floodwater.

The Nile might seem an ideal setting for village farming and a perfect, foreseeable environment for generating huge grain surpluses. The Greek historian Herodotus described *akhet* as a yearly event that appeared like clockwork. The popular myth of a reliable flood persists to this day, but the Nile is a capricious river. Exceptional rains meant potentially catastrophic inundations that carried everything before them and swept away crops and entire villages. A weak *akhet* covered only a small part of the river plain. Sometimes the flood receded almost immediately, crops failed, and hunger ensued. Most years, there was enough water for adequate crops; the farmers could survive short-term droughts without much trouble. But arid cycles of multiple years, decades, and even centuries were another matter.

ALL-POWERFUL PHARAOHS (3100 TO 2180 BCE)

Life was unpredictable; order and unity were required. For many centuries, Egypt had been a lattice of competing kingdoms, until a ruler (possibly) called Hor-Aha forged the Two Lands, Upper and Lower Egypt, into a unified state in 3100 BCE. Hor-Aha and his successors, who governed Egypt until 2118 BCE, presided over a state where the well-being of the common people depended on their supreme ruler, a terrestrial monarch whose reign epitomized the triumph of order over chaos. For nearly eight centuries, their terrestrial state functioned fairly smoothly.

Egypt was a civilization based not on dense city populations but on towns and villages linked by waterborne transport. This infrastructure kept the linear state together, with none of the logistical fifty-kilometer limits imposed by grain-laden animals. Fortunately for the pharaohs, the deserts that pressed on the Nile valley were natural fortifications, these and the shoal-laden delta making invasion nearly impossible. This contrasted sharply with the permeable and ever-changing borders of Mesopotamia and its two rivers, whose history involved the competitive rise, fall, and sometimes rise again of different kings and their cultural groups. Meanwhile, Egypt's natural isolation allowed the pharaohs to control their subjects. The population was organized but dispersed; censuses and taxation levied in grain, animals, and other commodities assured ample food surpluses; the state asserted its rights to prime agricultural land.

A pharaoh could easily dominate his circumscribed realm, provided that the government was seen as beneficial and powerful. Kingship was both eternal and personal, symbolized by the tangible divinity of the ruler. Egyptian kingship was an institution, marked by the success or failure of the pharaoh. Ultimately, for all the pharaoh's perceived divinity, royal authority depended on ample food surpluses that in turn depended on the toil of the people. For all the complexities of day-to-day political challenges and the sometimes-subversive deeds of provincial governors, the state was, above all, vulnerable to climatic shifts that brought severe droughts—prompted by weakened monsoons in the Indian Ocean,

The Old Kingdom pharaohs, who ruled between 2575 and about 2180 BCE, were powerful, confident rulers who presided over four centuries of good floods and bountiful crops. They could easily proclaim, with their godly status, that they controlled the inundation in all their divine majesty. The pharaoh ruled from his court at Memphis in Lower Egypt, twenty kilometers south of the Pyramids of Giza. He presided over the "unified" state of Upper and Lower Egypt, comprising nine nomes (provinces) ruled by powerful and fractious nomarchs (provincial governors). As long as the inundation brought enough water, the king's power was relatively secure. The leaders expanded irrigation works and canals and intensified agriculture in Lower Egypt's fertile delta. But a poor flood and crop failures undermined the most critical element of the state's power: adequate food surpluses. There were, of course, occasional poor flood years, but plentiful waters always returned. The state was powerful and successful, so much so that Egypt's population had risen to more than one million people by 2250 BCE, many of them dependent to some degree on the state for food.

Sometime after 2650 BCE, an increasingly powerful priesthood linked sun worship to the cult of the pharaoh. Upon the ruler's death, he would take his place among the stars, considered divine beings. "The king goes to his double . . . [A] ladder is set up for him that he may ascend on it," records a Pyramid Text inscribed in a pyramid chamber.[4] The pyramids built by the Old Kingdom pharaohs were stone symbols of the sun's rays bursting through the clouds. These imposing stone ladders boasted of the king's mortuary temple on their east sides, facing the rising sun. Their construction was a triumph of bureaucratic organization: arranging the transport of food rations and raw materials and marshaling skilled artisans and thousands of village laborers every flood season when agriculture was at a standstill and more labor was available. Everyone is familiar with the colossal Pyramids of Giza to the west of Cairo, built around 2500 BCE, but precisely why the pharaohs erected such elaborate and labor-intensive tombs remains a mystery.[5] Perhaps their purpose was to link the people, via the workforce, to their guardian. This would serve also as an administrative device to organize and institutionalize the relationship by redistributing food for labor—a technique that could be harnessed at times of scarcity. Or maybe the pyramids were built primarily to emphasize the extraordinary pharaonic connection to the gods, a method of connecting the king to the sun god, the ultimate source of human life and abundant harvests. We will never know. After a while the pyramids had fulfilled their purpose. State-directed labor shifted to other, less conspicuous projects.

A vast gulf separated the elite, including literate scribes, from hardworking commoners, required to provide labor to clear irrigation canals, haul stone, and grow crops. This was a time of well-directed, authoritarian leadership that depended on close, cooperative relationships between the pharaoh, his nomarchs, and high officials. Their collective talents and military force created a unique civilization that worked well in better-watered centuries but was highly vulnerable, indeed fragile, when *akhet* withdrew its bounty.

THE MEGADROUGHT STRIKES (C. 2200 TO 2184 BCE)

The vulnerability came home to roost immediately after the reign of the last great Old Kingdom pharaoh, Pepi II (2278 to 2184 BCE), who is said to have ruled for a staggering ninety-four years, the longest reign in Egyptian history.[6] As he became older and less effective, his nomarchs became restless. He responded by bestowing great wealth on the provincial governors, which effectively diluted his centralized authority. At his death in 2184, chaos ensued as high officials competed for power. This was the moment when the 4.2 ka Event that devastated Mesopotamia descended over the Nile.[7]

Numerous pieces of evidence document the increased aridity. At the Blue Nile's source, Lake Tana in Ethiopia, freshwater cores document a dry spell in 2200 BCE. In the Red Sea, brine sediments hint at a major drought at the same time. A core sunk at Saqqara in Lower Egypt revealed a meter of dune sand lying over formerly arable fields. Low floods, as well as occasional intense rainstorms, cut off Lake Qarun in the Fayum Depression from the Nile. Even tree rings lifted from a cedar coffin and a funerary boat show signs of a drought from 2200 to 1900 BCE.

A sudden and catastrophic long-term decline in the inundation caused almost immediate famine and paralyzed well-established political institutions. There were repeated famines for three hundred years, with many more mouths to feed than in earlier times. Desperate farmers planted their crops on river sandbanks, but to no avail. A sage named Ipuwer may have witnessed the drought at first hand. He described Upper Egypt as "a wasteland." "Lo, and all say, 'I wish I were dead.'" In a comment that resonates today, he blamed the pharaoh: "Authority, Knowledge and Truth are with you, yet confusion is what you set through the land, also the noise of tumult."[8]

People naturally turned to the pharaoh in Memphis for relief, as he had long proclaimed his mastery over the capricious river. Pepi's successors were incompetent and powerless. Stored grain soon ran short. A revolving door of rulers cycled through Memphis, while political and economic power passed to the nomes, now a patchwork of small kingdoms presided over by ambitious nomarchs, some of whom ruled like kings. Competent nomarchs took draconian steps to take care of their people. They soon learned one of the cardinal rules of coping with sudden climate change through practical experience: tackle the problem locally.

Some of the nomarchs were given to boasting of their achievements on their tomb walls. To what extent their boasts reflect opportunism rather than actual deeds is a matter of debate. Ankhtifi of Nekhen and Edfu ruled two of the southernmost Egyptian nomes in about 2180 BCE, at a time of exceptionally low Nile floods. His tomb inscriptions talk of his decisive actions: "All of Upper Egypt

The nomarch Ankhtifi depicted on the wall of his tomb. He was an effective administrator in the face of hunger and low Nile floods. *History and Art Collection/ Alamy Stock Photo.*

was dying of hunger, to such a degree that everyone had come to eating his children. But I managed that no one died of hunger in this nome."[9] Ankhtifi loaned precious grain to other provinces. His boastful, self-serving tomb inscriptions describe how people wandered aimlessly in search of food. Such behavior is eerily similar to that of people in India during the horrific famines of Victorian times in 1877 (see Chapter 9). Once prosperous nomes became arid wastes as sand dunes blew onto the floodplain from the surrounding deserts. Storehouses were empty; tomb robbers looted the dead.

Like Ankhtifi, the nomarch Khety of Assiut took drastic measures to combat famine. He built storage dams, drained swamps, and dug a ten-meter-wide canal to bring irrigation water to drought-stricken farmland. Competent officials knew that only drastic measures would feed everyone. They closed the boundaries of their provinces to prevent uncontrolled wandering by hungry people. They rationed grain and distributed it with sedulous care. Powerful nomarchs were the real rulers of Egypt, for they alone could take short- and longer-term measures to feed the hungry and stimulate local agriculture. The fragile unity of the Egyptian state collapsed.

For three centuries, Egypt was a fractured civilization. The pharaohs had fostered the belief that they controlled the mysterious inundation that arrived from far upstream. In fact, the Egyptian state in all its arrogant magnificence depended on the vagaries of Indian Ocean monsoons and atmospheric changes in the distant southwestern Pacific. The crisis finally ended with higher floods and hard-fought military campaigns by the pharaoh Mentuhotep, who came to the throne in Upper Egypt in 2060 BCE, reunified the state, and reigned for half a century.

Mentuhotep and his successors rebuilt the agricultural economy during reigns when the pharaoh was no longer perceived as infallible. They became "shepherds of the people" who imposed a firm bureaucracy on every aspect of Egyptian life. They were blessed with ample floods except during the eighth and seventh centuries BCE, when low flood levels again led to political confusion. But, by this time, the nomarchs depended on one another as never before for economic survival. In later times, the most successful Egyptian pharaohs prospered because they deployed people to turn the Nile valley into an organized oasis. When Rameses II (1279–1213 BCE) built his city of Pi-Ramesses, his canals were said to be the most impressive in all of Egypt: highly efficient, grand, elaborately ornamented structures that watered the entire region.

The pharaohs were nothing short of godlike managers of a centralized agricultural state, where investment in expanded irrigation schemes, technological advances, and large-scale food storage ensured the survival of the people in years of hunger and crisis. Religion was the system's ultimate source of control. Individual farmers who built canals for their own fields and crops were mindful that they had to build fairly—or face damnation. Confessions 33 and 34 of the "Negative Confession," the proclamation that the soul would make after death when it stood to be judged, required the soul to say that it had never obstructed water in or illegally cut into someone else's canal.

Ultimately, the land was as ready for crisis as it could be. Egypt's preparedness is even remembered in the biblical story of how Joseph and his family went to Egypt to flee the famine in Canaan, knowing that Egypt would have adequate grain supplies.

Despite their all-seeing power, the gods were unable to provide long-range monsoon forecasts. The priests did develop simple Nilometers over the centuries, ingenious scientific devices that measured the height of the inundation as the water rose. Few now survive, except some dating from after the seventh-century Muslim conquest of Egypt. Most pharaonic meters were under the control of temples. An important example survives on Elephantine Island opposite Aswan in Upper Egypt, the southernmost city in the state. Here the earliest flood measurements of the season could be taken. The meter, built before Roman times and restored by the Romans, is basically a well on the bank of the river constructed from close-fitting stones marked with different recorded flood levels. Long experience passed down by generations of observers enabled the priests to predict floods levels with surprising accuracy. This was invaluable information not only for farmers coping with irrigation works but also for tax collectors, who supervised the harvest with unremitting care. As the Greek geographer Strabo observed cynically, the better the flood, the greater the revenue.

It was no coincidence that Egyptian civilization flourished for another 2,000 years and eventually became the granary of Rome, as we'll discuss in the next chapter. But, even then, sudden climatic shifts caused prolonged droughts and resulting famines that killed thousands and affected the supply of grain to Rome and Constantinople.

THE INDUS: CITIES AND COUNTRYSIDE (C. 2600 TO 1700 BCE)

The fluctuations of the Indian monsoon affected the lives of millions of people—not only in the Nile valley and Mesopotamia but also in tropical Africa and perhaps also in South and Southeast Asia—among them the inhabitants of the Indus River valley and surrounding landscapes.

South Asia is hemmed in by tropical forest in the east, by mountains in the north, and by the Arabian Sea, the Indian Ocean, and the Bay of Bengal. The subcontinent developed its own cultural identity and distinctive civilizations marked by great diversity. The earliest was the Indus Civilization, one of the great early civilizations that flourished simultaneously alongside Mesopotamia and Egypt.[10] Still poorly known to the wider world, it was almost accidentally discovered in the Punjab by British and Indian archeologists of the 1920s. We now know that it flourished over a vast area of at least 800,000 square kilometers (roughly a quarter the size of western Europe), covering modern-day Pakistan and extending from today's Afghanistan to India. The Indus and now dried-up Saraswati River valleys were its cultural focus, but they were only part of a much larger, diverse, and dispersed society that extended over a very broad range of environments from highland Baluchistan and the Himalayan foothills across the lowlands of the Punjab and Sind to today's Mumbai.

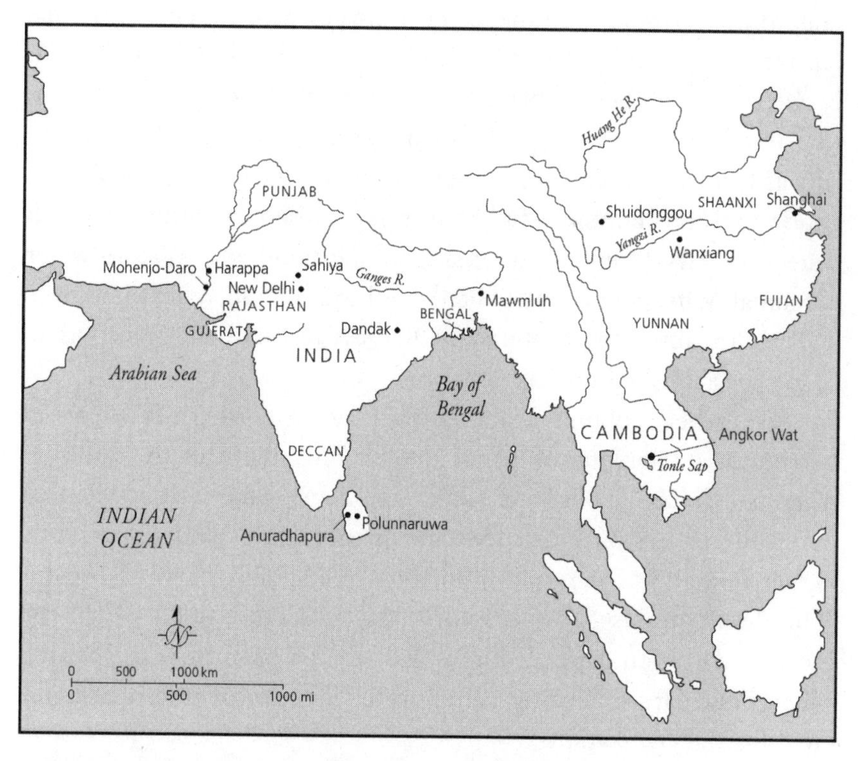

Sites in South Asia, Southeast Asia, and China.

Archeologists have identified well over 1,000 Indus settlements in multiple ecological zones, ranging from lush, green idylls to hot, inhospitable, semidesert areas. Though most of the sites were villages, at least five were major cities. To be clear, this was the largest urban culture of its time, roughly twice the size of its equivalents in Mesopotamia or Egypt. Its cities thrived for an impressive six to seven centuries between about 2600 and 1900 BCE. The population numbered perhaps one million, equivalent to ancient Rome at its height. Yet this massive civilization soon vanished from history. Neither the invading Alexander of the fourth century BCE nor the subcontinent's Buddhist-oriented ruler, Asoka, of the third century BCE was even vaguely aware of it. What role, ask the archeologists, did climate change play in the demise of the Indus?

Today, the local climate is good for agriculture since two different weather systems dominate, sometimes overlapping.[11] A rainy winter cyclonic system operates in the western highlands, while a summer monsoon system waters the peninsular regions. If one of these systems fails to deliver rain, the other almost always will, which means that famine is today unknown in the Indus valley. The Indus itself floods between July and September. Farmers plant their crops as the inundation recedes and harvest the following spring, using the flood-borne silts as fertilizer. Interestingly, there is no evidence of large-scale irrigation by Indus farmers—unlike in Egypt, where irrigation channels were necessary to extend the reach of the flood and to store water. Presumably, if a poor harvest befell one Indus region, relief came from another region with an abundant harvest through the transport of food via established trading networks.

Speleothem columns from Sahiya Cave in northern India, about two hundred kilometers north of New Delhi, tell us that the centuries when the Indus Civilization came into being were a time when an intensified monsoon created warmer temperatures and significantly higher rainfall.[12] The result: even more predictable harvests and reliable food surpluses, the economic superstructure upon which the Indus Civilization depended. This was when a patchwork of growing villages and larger farming communities developed into a complex preindustrial civilization.

Cities have become a hallmark of ancient civilizations, albeit in many guises. By no means all of them were the compact, crowded walled settlements found throughout much of the Middle East. Indus cities defy easy comparison to Uruk, Ur, Pi-Ramesses, or indeed any other cities elsewhere. Forget the grandiose pronouncements of Assyrian and Sumerian monarchs or the self-glorifying ideological proclamations of Egyptian pharaohs. The leaders who presided over Harappa, Mohenjo-Daro, and other Indus cities remain anonymous. Unlike the Egyptians or Mesopotamians, they were not given to advertising their achievements on temple walls. Then again, this civilization does not appear to have had temples: indeed, it lacks any overt signs of any religious structures. Moreover, there are only vague hints of any religion—such as the diminutive bust of a "priest-king," who may not be a king or a priest but simply someone in the act of blissful yogic-type meditation. Plentiful decorated seal-stamps also bear a wide variety of images, including people seated in apparent yoga positions. Religion? Maybe. Sadly, their writing system remains undeciphered. If ever cracked, the Indus code may tell a different story, but until then the archeology points to cities populated by self-effacing and egalitarian people.

The irrepressible British archeologist Mortimer Wheeler, who excavated at both Harappa and Mohenjo-Daro during the late 1940s, uncovered no richly adorned architecture, spectacular temples, gold-covered shrines, or palaces. Instead, he found two citadels with rather pragmatic public buildings, including a granary and a great pillared hall built of brick—brick offering protection from flooding. The people lived in precisely built (again, brick) houses that showed none of the usual urban signs of class differences. And yet, for all their apparent egalitarianism, both cities were among the most sophisticated in the world during their occupation between about 2550 and 1850 BCE. They had imposing flood-protection works, wells, and sanitation comparable to that of the modern era, including shower cubicles and plumbed toilets—the earliest in the world. At each city, the builders followed an irregular netlike plan that evolved over many generations and included grids of streets and houses. Wheeler memorably described his impressions: "Middle-class prosperity with zealous municipal supervision."[13]

A well at Lothal, India. *Dinodia Photos/Shutterstock.*

Wheeler rejoiced in vivid descriptions, colored by his Western perspective. But his depiction of middle-class prosperity is wrong. The latest interpretations talk of the cities as polycentric communities with walls and platforms delineating different zones within cities and lesser settlements outside the cities, where economic activities unfolded and artisans labored. Indus Civilization may have been a nonhierarchical society where communal activities were commonplace. Yet its city dwellers may also have been quarrelsome, as settled humans often are, and there are hints of rival local communities at Harappa, for example.[14] However, drawing attention to possible local spats is rather unfair of us, given that this is the only known civilization in the world with zero evidence of any organized warfare. Despite attempts to search for the opposite, all the evidence points to a society that was, at least at the city level, peaceful, prosperous, and egalitarian in outlook. It was well connected too: its people traded with the Persian Gulf and Mesopotamia for many centuries.

Whatever kind of society flourished along the Indus and beyond, it was certainly not a social pyramid. A greater contrast with the bombastic Egyptian and Mesopotamian states is hard to find—and in terms of combatting climate change, it was far more resilient, though fragile in the sense that rivalries between local leaders and cities were always in the background.

As the cities grew, so did the surrounding rural settlements. In fact, maybe we should call the cities "city-states" to reflect their importance in the local landscape. As to their satellite settlements, many of them centered around farming; others were craft centers. Many were occupied for a short time or discontinuously. Such residential instability was commonplace, especially in areas with braided river systems and frequent monsoon flooding. Environments like these required settled populations to be mobile as a way of adapting to rapidly changing hydrology. Part of this adjustment would have involved family members and kin spreading their numbers between several settlements as an equable way of accessing water supplies. These realities would have provided additional resilience and sustainability in what was, in places, a very challenging landscape.

Under these circumstances, risk mitigation was central to survival and probably included such strategies as multicropping (growing two or three crops annually), using drought-tolerant crops, and growing different cereals simultaneously on the same land.[15]

Agricultural diversification increased over time, with greater use of both winter crops, like barley and wheat, and summer crops, such as millet and drought-resistant grains. Farming practices varied widely over different areas, which would have made any form of centralized storage and control challenging. A large granary at Harappa shows that the feeding of sizable nonfarming urban populations was certainly a concern. Most likely, cities like Harappa relied on food surpluses from their immediate hinterlands, as well as well-established trade networks for basic commodities, while rural villages were basically self-sufficient.

The Indus Civilization was a stark contrast to the Egyptian one. This was no single unified state but a richly diverse, decentralized society, which made issues of sustainability infinitely more local concerns than they would have been if authoritarian lords presided over huge domains. Risk management varied greatly from one region to another, with one common development: all the Indus cities dissolved around 2000 to 1900 BCE, and with them the whole cultural complex. Why?

SURVIVING THE MEGADROUGHT

The 4.2 ka Event, a period of intense aridity, haunted both simple and more complex societies throughout Asia and the Indian Ocean world. The weakening of the Indian summer and winter monsoons coincided broadly with the dissolution of Harappa, Mohenjo-Daro, and other Indus cities, but it seems unlikely that the megadrought was the only trigger. We use the word "dissolve" on purpose, for it would be misleading to talk of collapse. There had been a long tradition of dispersal among rural communities. Recent isotopic research on skeletons from a Harappa cemetery reveals that many of the dead were immigrants from elsewhere. A constant flow of people entered and left the cities, as they did smaller communities. This is hardly surprising, given the close ties between villages and larger communities, which must have involved fellow kin or at least trading partners.

The dissolution of the Indus cities may have been just a defensive response to food shortages that could be met by moving to better-watered communities where food was available. This was a decentralized civilization, so mobility was adaptive. After all, why provide food to a city when you are better off looking after your own community? Adapting to long-term drought in the villages was a routine matter of diversifying crops and adopting more summer and drought-resistant grains such as millet and rice. The yields would have been lower, making it hard to support large cities. There were obviously variations across the Indus region, but, once again, one makes a distinction between short-term droughts and longer arid cycles, when short and even medium supply networks would have been unable to produce adequate city food surpluses. An ancient strategy of resilience came into decisive play in a heterarchical society that placed a high importance on kin links and obligations. According to settlement studies, many people left the Indus valley around 1800 BCE and moved northeast to Rajasthan and Harayana, which saw a major population increase as Harappa declined.

Resilience apart, fundamental questions remain. What happened to the seemingly robust Indus cities in the face of lengthy drought? Was the climate now too dry? Did the farmers' adaptation become too diverse? Did climate changes make it impossible for urban Indus populations to adapt? We know that while the Indus still flowed fast, the region's second major river, the Saraswati, dried up—perhaps the result of an earthquake that captured its headwaters and redirected them toward the Ganges. With the drying of the Saraswati River, the settlements that depended upon it also disappeared. This ultimately led to the upending of the entire society.

Despite its disappearance, in the great scheme of things, this had been a long-lived civilization. No question that the Indus cities were unusually robust and long-lasting by early preindustrial standards. Their long-term resilience may have resulted from reliance on sustainable rural lifeways that proved inadequate when reduced crop yields decimated food surpluses. In contrast, rural farmers achieved long-term sustainability by growing a range of crops adapted to their local environments and water supplies. Smaller populations would have had familiar, long used social mechanisms where choices about crops and farming practices as well as cultural behavior were more flexible. Under these circumstances, population displacement would have been essential in many places—which accounts for the constant settlement abandonment. Certainly, there is no evidence of a traumatic end to the civilization: we have no suggestion of any great wars (or even small ones) and no evidence of violence or destruction at the settlements.

The Indus Civilization was robust because it was based on a rural social and economic underpinning that was, by its very nature, resilient and sustainable, in part because the environment was so challenging and diverse and perhaps also because of the seemingly peaceful ideology, with its lack of social hierarchy or constraining religious dogma. This worked well in a decentralized society where much social authority remained local. Cities were temporary adaptations. Rural communities could survive long-term droughts, perhaps buffered by help from nearby communities, but certainly without the trauma experienced by hungry, dense urban populations. Again, the most successful climatic adaptions were ultimately local.

DIFFERENT STROKES

Fractious, fragile, and vulnerable: the earliest civilizations in Mesopotamia and Egypt were each a litany of rulers attempting to impose their wills and their distinctive modes of governance on what had always been village societies. Legitimized by their religions and their gods, theirs was a tale of power and the glory. In the Indus realm, the people seem to have tried something else: cooperation and social egalitarianism (at least among the city dwellers), with an apparent de-emphasis of hierarchy, monarchy, and religion. Each of these strategies of adaption to climate change succeeded for a while, until new systems of political organization emerged and transformed societies. Yet, when it came to dealing with drought and major climatic events, the most effective responses came not from centralized empires that conquered neighbors for their resources or powerful satraps who administered centralized storage warehouses but from local initiatives, which tailored adaptive responses to the familiar realities of their own communities and their surrounding environments. Without question, the same is true today.

None of these early civilizations was completely powerless in the face of climate change. Nor were they capable of infinite adaptability, facing, as they sometimes did, major happenings like the 4.2 ka Event. Their experience of biblically lengthy droughts is something that modern industrial civilizations have never had to confront. Just to put the arid event of 4,200 years ago in a contemporary context, the fifteen-year drought in the Levant from 1998 to 2012 is said to have been drier than any comparable period over the past nine hundred years. This drought was far greater than others caused by natural variability in recent centuries. The culprit was inexorable anthropogenic climate change. Given the predicted global climate of the future, we will need to make even greater adaptations on an international scale, far more than was the case in the past. Lessons from the 2200 BCE megadrought event may help us face the impending, massive climatic challenges of the future.

The legacies of these societies have considerable relevance today. The pharaohs ruled over a great river valley with little rainfall but an unpredictable, annual inundation. The 4.2 ka Event taught them that neither dictatorial power nor the gods would provide solutions to crop failure and starvation in a society where agricultural authority resided, ultimately, in the village. Later rulers promoted new doctrines of the pharaoh as a guiding shepherd. These leaders invested heavily in grain storage and locally based irrigation schemes. Their civilization endured for more than 2,000 years. Meanwhile, over in Mesopotamia, its people dwelt in a fractured political landscape, defined in considerable part by dramatic climatic extremes and often violent floods. The environment was far more variable than that of Egypt and one in which volatile environmental changes could cause rivers to shift their courses or even dry up. In the long term, survival and adaptation to drought cycles and other climatic variables required both intimate environmental knowledge and profound farming expertise. Here, the real power lay ultimately not in the hands of powerful, conquering kings but in the ability of cities and farming communities to adapt to local conditions. As the Sasanians found to their cost centuries after the Assyrians, large-scale irrigation agriculture, with its sweeping environmental changes, was vulnerable in ways—notably to salinization—that were already known to smaller-scale farmers of earlier times. Sasanian agriculture failed.

Along the Nile and in Mesopotamia, an elite minority ruled. They lived in golden luxury while the peasants toiled, sometimes in chronic poverty. For the few to control the many, centralized political and economic control was the ideal—even if it meant suppressing local knowledge, traditional solutions, and resilience in the face of ever-growing demands for taxes in kind. The Indus Civilization appears to have been the opposite: a decentralized, very diverse society that promoted social equality (at least in its cities) and in which power lay in small communities near the soil. Here mobility was a frequent adaption to failed inundations and drought. Even when the Saraswati River dried up and the Indus cities dissolved, the distinctive Indus culture and its institutions survived for a time. If any example from the past proclaims the value of traditional knowledge and local solutions to coping with climate change, the Indus Civilization is it.

Our modern industrialized world, meanwhile, operates a system of extreme economic inequality, built on an ideology of accumulation, growth, and exploitation, in which an elite few grow rich on the labor of others. Yet many capitalists forget, or prefer to ignore, the countless people who live rurally and follow more traditional lifestyles. Such people survive, albeit often with difficultly, because they rely on ancient, traditional farming and herding strategies that, crucially for all our futures, are still sustainable in the modern world.

While archeologists have taught us a lot about climate change and adaptation in very ancient times, we have many historical accounts and scientific data spanning the past 2,000 years. As we will see, even short droughts of a few decades or brief cold snaps contributed to death and misery and, ultimately, to the fall of the mightiest empires. In the following chapters, we begin in Italy and then wind our way across the world, exploring several other empires that imploded in the face of climate change. At times, we will see how people dealt successfully with climatic challenges, and we'll learn from them. But first, the fate of Rome.

5

THE FALL OF ROME

(c. 200 BCE to the Eighth Century CE)

Rome: the very scale of the empire at its height in 350 CE boggles the mind. Rome's citizens prospered in Spain at the western tip of Europe and as far east as the Nile valley. Her legions garrisoned Hadrian's Wall in chilly northern Britain, manned fortifications along the Rhine and Danube, and were a significant presence on the northern margins of the Sahara Desert and in western Asia. The Eternal City itself had begun as a small town, founded according to legend in 753 BCE by twin sons Romulus and Remus, said to have been raised by a she-wolf. Rome became a monarchy, then a republic, and finally the center of a huge empire. Yet, in 476 CE, it collapsed after the abdication of the last emperor.

Why the Roman Empire disintegrated is one of the great controversies of history.[1] In 1984, the German classical scholar Alexander Demandt described no fewer than 210 causes for its fall proposed since late classical times. Doubtless there are many more today, but with one major difference: we now know much more about climate change and its impact on life during Roman times.

BALMY BEGINNINGS (C. 200 BCE TO 150 CE)

The Roman Empire came into being during a period of warm, generally wet, and stable climate, conventionally named the Roman Climatic Optimum (RCO), which lasted from around 200 BCE until

150 CE.[2] These benign conditions coincided with much reduced volcanic activity after a major eruption of Alaska's Okmok II volcano in 43 BCE. No major eruptions occurred between the assassination of Julius Caesar in 44 BCE and 169 CE (even the famous 79 CE eruption of Vesuvius was relatively minor). In the West, the North Atlantic Oscillation (NAO) and Atlantic westerlies were dominant players. In the East, concatenated climatic actors were involved, among them the Indian Ocean monsoon, El Niños, and a persistent subtropical high at thirty degrees north, which suppressed rainfall with monotonous regularity. This was a time of warmth and climatic stability: perfect conditions for any *Homo sapiens*. Forty-five Alpine glaciers retreated until the third century. High-altitude tree rings show us that the highest temperatures were in the mid–first century. None other than the contemporary Roman naturalist Pliny the Elder remarked that beech trees were loving the mountains instead of merely flourishing at lower elevations. There was consistent humidity and good rainfall across the Mediterranean region.

The RCO, with its warmth and usually adequate rains, worked miracles for Mediterranean agriculture, especially wheat, which is very sensitive to rainfall and temperature changes. Years of greater warmth and ample precipitation extended the limits of cultivation and amplified the productivity of the land, to the point that cereal agriculture was more fecund in Roman times than it was centuries later in the hands of medieval farmers. One conservative estimate calculates that an extended temperature rise of 1°C made an additional million hectares of land suitable for arable farming—enough to feed an additional three to four million people. Not only did wheat farming expand, but staples such as olives and grape vines did as well.

Three powerful factors acted in unison to foster the expansion of Rome's domains: trade, technology, and climate. Increased rainfall turned North Africa into one of Rome's granaries. Today, North African countries import grain. Growing population densities pushed farmers into more marginal landscapes. As the empire grew and stabilized, the level of connectivity and long-distance trade improved dramatically, making risky farming a more realistic, less hazardous activity. Semiarid North Africa witnessed an explosion in irrigation agriculture with the construction of aqueducts, dams, cisterns, and the

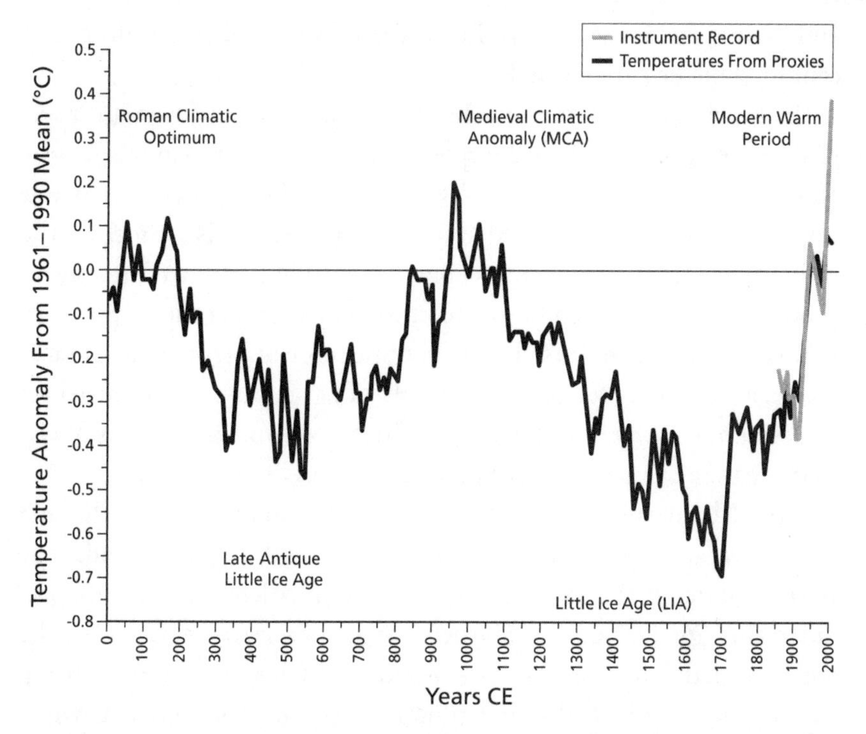

A highly generalized diagram of temperature changes in the European region over the past 2,000 years. Much more detailed charts can be found in the references cited in the notes.

simple but ingenious *foggara* that used gravity to transport groundwater from higher elevations to cultivable lowlands.[3] At the height of the RCO, cultivation extended into what today is the northern Sahara Desert. The desert began to advance once more during the second century, when aridity returned. To the east, speleothems from Soreq Cave in the Dead Sea region have recorded a sharp decline in rainfall after 100 CE.

An inexorable shift toward greater summer aridity accelerated toward the end of the RCO. One scenario argues that this was the result of Roman farmers stripping Mediterranean forests for wood for construction, fire, and fuel. Such activity would cause more heat to be reflected from the ground. Thus, there was less soil evaporation into the lower atmosphere, resulting in lower precipitation in summer. If this scenario is correct—the debate continues—then anthropogenic and natural factors came together as the RCO ended, and centuries of stress for the empire ensued.

"The climate was the enabling background of the Roman miracle," argues classicist Kyle Harper.[4] He describes the lands ruled by Rome as "a giant greenhouse." The RCO produced growth that was unprecedented in its scale and ambition. But—and it is a big but—the stability of this seemingly miraculous expansion depended directly on powerful factors beyond human control. For three centuries after 150 CE, the Roman Empire's climate became more unpredictable and unstable, a capricious reworking not only of agriculture and governance but of demography into the bargain. The uncontrollable powers of climate change came into subtle, and sometimes dramatic, play.

The Mediterranean has always been a region of dramatic climatic variations, and the RCO, with its warmth and ample rainfall, may have moderated the excesses of annual unpredictability, an important reality for farmers, as Harper further observes. In 128 CE, the well-traveled Emperor Hadrian visited his African provinces. During his visit, it rained for the first time in five years during a year when wheat prices were 25 percent higher than in wetter decades. The miracle of imperial rainfall was all very well, but practical steps were needed. Hadrian boldly ordered the construction of a 120-kilometer-long aqueduct, one of the longest the Romans ever built, to bring water to Carthage.[5] The emperor's act of resilience was remarkable but merely one response to an enduring aridity crisis that descended over the heart of the empire over several centuries.

RESILIENCE AND PANDEMICS (FIRST CENTURY CE ONWARD)

The Roman Empire was a huge tapestry of agricultural, demographic, fiscal, military, and political systems. All kinds of risks imperiled the state. As Emperor Marcus Aurelius put it, the empire was like a windswept island besieged by hostile fleets, pirates, and storms. Every emperor had to confront adversity in a world of constant turbulence, including climate change. Risk management depended on human beings, using hard-learned strategies to confront such events as unexpected floods, prolonged droughts, and resulting famines that taxed granary supplies, and so on. Stress, much of it increasingly due to climate change, was a constant reality in the later Roman Empire.

The most effective weapon was in the countryside, among farming communities that had acquired inherited experience and expertise over many generations: crop diversification, conservative storage strategies, and exotic local crops that flourished in dry years and were valuable insurance. Self-sufficiency, reciprocity where one helped one's suffering kin and neighbors in hungry times, and delicately arranged patronage were all part of the farmer's armory. A profound spirit of independence lay behind rural societies in the empire. In Britain, for example, Roman-era farming settlements appear to have exercised a degree of autonomy. While only a few such sites are known, two have been excavated in Somerset in southern England. The first, Sigwells, consists of isolated stone-walled rectangular buildings, whereas nearby Catsgore is marked by a linear "street" arrangement.[6] The two Romano-British settlements are contemporary yet look entirely different. They were clearly not blanket-organized according to top-down imperial rules but developed individually, according to the local needs of the occupants, and sometimes—as in the case of Sigwells—over very long, pre-Roman time spans.

Strategies of resilience extended into cities and towns. Urban food storage assumed great importance throughout the empire. It was no coincidence that many cities developed along major rivers and waterways, which reduced their dependence on their own hinterlands. Inland cities were notoriously vulnerable to short-term droughts because importing and exporting food was much harder.

When food crises unfolded, the Roman government was prepared and either provided grain or suppressed any attempts at exploitation. In a real sense, this was an extension of the principles of reciprocity and patronage that were commonplace in the countryside. Imperial strategies were often on a grand scale. During his 117–138 CE reign, Emperor Hadrian visited many cities and "took care of them all."[7] He provided water with aqueducts, built harbors, imported grain, and even gave money for constructing public works. Municipal granaries such as those that fed Rome were on a monumental scale. Emperor Septimius Severus (r. 193–211 CE) was so concerned about feeding Rome that he left seven years of grain at his death. Grain doles became an established symbol of imperial generosity. An official letter of the second century from the ancient city of Ephesus promises grain from Egypt provided the harvest yield was adequate to feed Rome. "If, as we pray, the Nile provides us with a flood of the customary level and a bountiful harvest of wheat is produced among the Egyptians, then you will be among the first after the homeland."[8] In many ways, the Romans had the same challenge with global food supplies as we do today, for the vulnerability to hunger was ever growing. Contemplate modern American or European supermarkets. You can buy food from six continents. Like the Romans, our food supplies also depend heavily on monoculture, on large-scale production of maize, wheat, and other cereals, and also on industrial animal husbandry. What would happen if parts of the human food chain were to collapse due to global warming? Or what about the impact on food supplies in the face of human pandemics like Covid-19 and also animal epidemics, such as "mad cow disease," which could decimate beef supplies in short order?

The Roman food chains achieved significant complexity. During the second century, about 200,000 citizens of Rome received five *modii* of wheat monthly, about 80,000 tons for the dole alone.[9] A flotilla of large grain ships sailed each year from Alexandria to Rome, always welcomed by joyous crowds. Remarkably, the transport of grain to the city was left in private hands without official involvement, a tribute to the great strength of the grain market. But most of Rome's cereal supply depended on two major granaries: the provinces in North Africa and in Egypt.

Throughout its imperial history, the Roman Empire was a vast enterprise based on growing cities and commercial networks that extended far beyond its borders. The Romans were well aware of the Chinese. Imperial Rome was a civilization of grandeur and awe, which fostered movement and connectivity over enormous distances. It also became an incubator of pandemics, in considerable part because of urban sanitation issues. The major cities of the empire were densely populated, with people living at close quarters, crowded with immigrants and slaves from distant lands. Roman civil engineers brought water to the hearts of cities for drinking and bathing and for flushing sewers. The larger public toilets they constructed could cater to between fifty and a hundred clients at one time. But waste disposal and sanitation were at best rudimentary. Rome alone is said to have produced over 45,300 kilograms of human excrement daily. Roundworms and tapeworms, along with other parasites, were commonplace, and a massive germ pool made cities lethal cocktails of infection, especially during the peak death period in the late summer and fall, when summer heat was a killer. Malaria, typhoid, chronic salmonella, and diarrhea affected affluent and poor alike. Even the emperor was not immune: Titus likely succumbed to malaria in 81 CE. The warmer centuries of the RCO, with the increased rainfall, seem to have fostered epidemics of malaria. Rome and other major cities were petri dishes of infection.

The epidemics generally came from within rather than from outside, until the time of Marcus Aurelius during the second century CE, when commercial links with the Indian Ocean and the Bay of Bengal increased dramatically, thanks to Rome's exploitation of the monsoon wind sailing routes.[10] By this time, some 120 merchant vessels from India arrived annually in Red Sea ports. They brought gold, ivory, pepper, and other spices, in addition to Chinese silk. Pepper became a staple, even for soldiers serving as far away as Hadrian's Wall in northern Britain. Alexandria, situated at the crossroads of the Mediterranean and the Indian Ocean world, became the greatest market for eastern luxuries. Much of the trade originated on the ivory- and gold-rich East African coast—a region of great microbial biodiversity and potentially lethal pathogens for humans.

The Silk Roads that crossed Eurasia were also long-established routes for human-incubated pathogens. In 2016, researchers working on a large "Silk Road relay station" in northwestern China found the oldest-known evidence for travelers spreading infectious disease over vast distances. Their work centered on a Han Dynasty latrine dug around 111 BCE and still in use until 109 CE. On bundles of "personal hygiene sticks" (i.e., fabric-wrapped sticks used to wipe away feces), the team found eggs from four different species of parasitic worm—including those of the Chinese liver fluke, a parasitic flatworm that causes abdominal pain, diarrhea, jaundice, and liver cancer.[11] The worm requires well-watered, marshy areas to complete its life cycle; yet the Xuanquanzhi relay station lies at the eastern end of the arid Tarim basin. This means that the liver fluke could not have been common to this dry region, and the closest endemic area today is around 1,500 kilometers away. The conclusion: an individual traveler, preinfected with contagious liver fluke, must have journeyed a huge distance, bringing his bellyache with him. But worms and their eggs were nothing compared to what would soon beset the world.

In the mid-second century, a fast-moving pestilence that seems to have originated in tropical Africa spread into Arabia, perhaps in 156 CE, during the reign of Antoninus Pius. By late 166, what is now known as the Antonine Plague had arrived in Rome and spread rapidly through the western Mediterranean from one concentration of people to another.[12] Entire legions were decimated; recruitment plummeted. The pandemic, the first recorded in history, spread from southeast to northwest, moving unpredictably. How many people perished is impossible to estimate, but it could have been as many as a third of the imperial population. The eminent Roman physician Galen described symptoms that most closely resemble those of smallpox, a disease transmitted by contact between individuals. In large cities like Alexandria, the disease lurked, then exploded. A major outbreak in Rome in 191 CE killed over 2,000 people a day. The Antonine Plague ricocheted across the empire at a critical moment when international trade connections had developed a new maturity.

The empire did not collapse, despite major economic disruption and population loss, for the next major plague outbreak only arrived in 249 CE. Population levels soon recovered, and the Antonine Plague left no permanent demographic effect. The main, short-term consequences were interruptions in basic food production and agriculture, with widespread famine in outlying areas of the empire. In some places, city dwellers descended on the countryside and stripped rural communities of food supplies that they felt were theirs. There were major political adjustments that need not concern us here, but unpredictable climate change and new germs that lurked over the horizon had exposed the vulnerability of the empire.

The smallpox outbreaks and persistent droughts caused widespread pessimism. By the late 240s CE, Cyprian, bishop of Carthage, writing in an increasingly arid North Africa, complained, "The world has grown old and does not stand in the vigor whereby it once stood . . . Come winter there is not such an abundance of rains to nourish the seeds. The summer sun burns less bright over the fields of grain."[13] He thought of the world as a pale old man near death, but he was wrong.

LOGISTICS AND VULNERABILITY (FOURTH CENTURY CE)

Despite Cyprian's pessimism, the empire prospered for most of the fourth century. Rome still had a special aura. About 700,000 people dwelt there, given a daily ration of baked bread—rather than grain—olive oil, and wine at a fraction of market price;[14] 120,000 people received pork handouts. All these free rations swelled the capital's population dramatically. At the center of everything was a vast military complex run by the state. Half a million men were in the field. A sophisticated logistical system provided every item of equipment, cavalry mounts and pack animals, and food. The food needs of the army alone were a huge burden on the empire, making it vulnerable to drought and other climatic shifts that were more severe than imperial authorities realized. Meanwhile, Constantinople (now Istanbul), founded in 330 CE, became the center of the growing Eastern Roman Empire. The city's population expanded tenfold from about 30,000 to 300,000 people during the fourth century. Grain destined for Rome now traveled east. As Kyle Harper aptly remarks, "So many ships covered the sea between Alexandria and Constantinople that it was like a long artificial strip of 'dry land.'"[15] The fourth-century city was a crossroads of international trade and a great center of Greek culture.

Fortunately, the climate was still relatively benign and warmer, encouraging economic growth, but the halcyon days of the Roman Climatic Optimum never returned. For all its prosperity, the empire depended on intensive monoculture and, above all, on imported grain from North Africa. Even in dry years, the most reliable food source was the Nile valley with its seemingly bountiful inundation, which originated from monsoon rains. A combination of fertile floodplain and lavish floodwaters created a natural irrigation system that people had modified and used since before the pharaohs. Egypt subsequently fed Rome and much of the empire.

Yet even Egypt's ingenious Nilometers could not predict the inexorable, long-term climatic changes that governed the Nile flood. The ultimate villains were the Intertropical Convergence Zone and the Indian Ocean monsoon far to the south and east, which shifted ever so gradually southward. The Nile floods were either stable or erratic, which affected human societies and civilizations along the river in dramatic ways. Meticulous studies of papyri have shown that when Octavian (later Emperor Augustus) annexed Egypt in 30 BCE, he did so during a period of dependable inundations and many excellent floods, which lasted until 155 CE. From 156 onward, the inundation became much less reliable, and food exports from the once fertile Egypt suffered, often dramatically.

Apart from monsoon shifts, a high North Atlantic Oscillation brought unpredictable conditions.[16] A long period of positive NAO conditions had begun in the late third century and lasted throughout the fourth at levels only seen later during the Medieval Climate Anomaly (see Chapter 11). Alpine glaciers retreated. In Britain tree rings chronicle high rainfall levels in northern and central Europe. French and German oak tree rings record rising quantities of rainfall until the early fifth century. But the period of good rainfall did not last. Three centuries of less stable climatic conditions followed. Beryllium isotope records show a major drop in solar insolation—the amount of sunlight reaching the earth. Cooling ensued; Alpine glaciers advanced once more. Major drought descended on the southern rim of the Mediterranean, devastating North Africa. Cities became short of food, while the rich sought to profiteer from rising grain prices. The rains failed along the Levantine coast, long known for its unpredictable precipitation. More plentiful downpours arrived just in time, but the stories of the great drought long survived in Jewish rabbinical writings.

The edges of winter storm tracks flickered over the Mediterranean, while tropical monsoons and distant El Niños caused flutters in rainfall over the eastern empire. Droughts and famines were more frequent. In 383 CE, harvests were grossly inadequate throughout many provinces, as the Nile flood was extremely low. The general scarcity of food was so severe that neighboring provinces were unable to help one another with grain shipments as they had done in the past.

For centuries, Roman philosophers and poets had written of a calm, benevolent universe. But now malevolent forces had descended on humanity. Predictably, people believed that either the Christian God in the newly Christianized empire of the fourth century or the pagan gods of those in the provinces who had not converted were angry and withholding rainfall. Inevitably epidemic diseases accompanied famine outbreaks, partly because people ate virtually inedible or toxic foods, which lowered their resistance to infections of all kinds.

HORSES, HUNS, AND HORRORS (C. 370 TO C. 450 CE)

To the east of the western empire lay the vast Eurasian steppe, a treeless expanse of grasslands and scrub. Rainfall was irregular and unpredictable, dependent on storm tracks from the west. The Romans despised the nomadic pastoralists who roamed the steppe where farming was impossible. While both the Romans and the Han in China were settled farmers, the nomads were constantly on the move, riding horses and herding cattle, while pressing against the settled lands, at first in China, then later in the west. During the fourth century, bands of nomadic Huns appeared at the frontiers of Roman power in the west. A sequence of juniper tree rings from the Tibetan plateau come from an environment in which the continental weather patterns and monsoons mix. From about 350 to 370 CE, the worst period of megadrought in 2,000 years affected the region. This may have triggered the Hun groups to move westward.[17]

The effect that sucked people in and out of arid environments—in during periods of more abundant rainfall, out during droughts—came into play. The Huns responded to the drought by jumping on their horses and fanning out in search of better-watered pastures for their herds. The center of political power on the steppe shifted from the Altai region of Siberia to the west. This sudden movement coincided with a period of intense competition between different alliances of nomadic groups. The Roman soldier and historian Ammianus Marcellinus described the Huns in graphic terms: "Although they have the form of men, however ugly, they are so hardy in their mode of life that they have no need of fire nor of savory food . . . They are almost glued to their horses."[18] Their powerful reflex bows were said to have a range of 150 meters. The sheer ferocity of their tactics was terrifying.

The Hun situation came to a head as nomads moved westward from the Middle Danube region. Emperor Valens was defeated in a sanguinary battle near the city of Adrianople in 378 CE. As many as 20,000 Romans perished in the slaughter. Between 405 and 410 CE, the western empire evaporated in the face of invasions by Goths and later other groups, who crossed the Rhine, ravaged Gaul, and conquered as far west as Spain. After the death of Emperor Theodosius I in 395 CE, the eastern and western halves of the empire were never again ruled by a single leader. In 410, the Gothic ruler Alaric entered Rome. The western empire had no military power left, and Rome's power was fractured. Attila, the most notorious of all Hun leaders, plundered the Balkans. Only a plague epidemic halted him at the gates of Constantinople, which had been ravaged by a massive earthquake in 447. Attila then advanced into Gaul and Italy but retreated to the steppes in the face of famine and an epidemic of malaria contracted in humid lowlands. By the sixth century, Rome's population collapsed, dependent as it had been on grain from elsewhere.

Emperors Diocletian and Constantine had tightened the control of the imperial administration during the early fourth century. They proclaimed themselves divine rulers who thrived on the prosperity of the eastern provinces. Diocletian transformed the emperor into a remote king, who relied heavily on ceremonial statecraft to extend his power in contrast to earlier emperors who moved from city to city. Constantine founded his capital on the ocean, on trade routes connecting east and west. His reign was the basis of the late Roman Empire. Constantinople replaced Rome as the crossroads of international trade and a great center of Greek culture. Grain destined for Rome was now diverted east.

Nowhere was imperial power displayed so prominently as with the annual audit of the imperial grain stores. Ultimately, the emperor's most basic obligation was to feed his subjects. With half a million inhabitants in his capital, he could leave nothing to chance. A vast bureaucracy controlled taxes and food supplies. At stake was the security of the city, achieved by supplying food. The threat of famine that had caused civil disorder in Rome led to vast stocks of grain stored to feed half a million people; 80,000 of them alone received free bread rations. Constantinople's food supply came from Egypt, as it had for centuries. During Justinian's reign (527–565 CE), 310,000 liters of wheat arrived by ship from Alexandria annually.[19]

Each year the emperor would mount his chariot. The praetorian prefect, the second-most powerful individual in the empire, kissed his feet. The imperial procession entered the busy market district of the city and then advanced to the huge public storehouses on the Golden Horn, where ships anchored with their cargos. Here the director of the granaries presented his accounts. If all was well, the director and his accountant were rewarded with ten pounds of gold and silk tunics. This elaborate, carefully staged public spectacle showed everyone that food supplies were secure.

Justinian presided over a truly global, volatile city crowded with people and goods from all corners of the known world. Constantinople was a cosmopolitan metropolis at the center of a vast network of lesser cities. But as the emperor and his courtiers toured the storehouses, another member of the ecosystem watched, invisible: the black rat, *Rattus rattus*. This ubiquitous rodent carried *Yersinia pestis*, the microbe that caused bubonic plague.

The plague arrived in Egypt in 541 and spread throughout the Roman Empire during the next two centuries. What became known as the Justinian Plague originated in the highlands of western China.[20] By the sixth century, trade with Asia, both overland and across the ancient Indian Ocean trade routes, was big business, especially in pepper and other spices. Silk was also a precious commodity, much of its production centered in the Red Sea. To the west of the Red Sea was the Christian Axumite kingdom of Ethiopia, and to the east was the Himyarite kingdom of southern Arabia, which at the time espoused Judaism and allied variously with Rome and Persia. This region was hugely strategic. It's little wonder that Mohammad, the prophet of Islam, decided to be born on the Red Sea side of Arabia, in Mecca, in 571 CE.

Germs followed the merchants, as did black rats infected with plague hiding in ships' cargos. The epidemic first appeared at Pelusium, close to the northern Red Sea port of Clysma, where ships from India docked regularly. From there, the plague traveled comfortably to the Nile and then into the Roman Empire. Once ashore, it spread in two directions—one west to Alexandria and then up the Nile valley, the other east, enveloping not only the Mediterranean coast but all of Syria and Mesopotamia. Efficient Roman networks carried the plague inland, most rapidly by sea. The pandemic arrived in Constantinople in March 1542 and lasted two months. At its peak, 16,000 people are said to have perished daily. Between 250,000 and 300,000 of the city's half-million population died. Local society collapsed, markets were closed, and famine resulted. Officialdom was decimated. Corpses piled up everywhere, despite mass burials in large pits. Many of the dead lay in layers, sinking into the "pus of those below." The churchman John of Ephesus observed the horror at first hand and wrote that he was watching "the wine press of the fury of the wrath of God."[21] The state reeled under the catastrophe. The price of wheat collapsed, for there were far fewer mouths to feed. An acute fiscal crisis undermined the state, which could barely mobilize, let alone pay, an army. Demographic collapse in the eastern empire was on the horizon. Between 542 and 619, plague struck Constantinople on average every 15.4 years. In 747, so many people died in a renewed plague attack that the emperor repopulated the nearly deserted city by forced migration.

FRIGID TIMES (450 TO C. 700 CE)

At this critical moment in Rome's history, between 450 and around 700 CE, three centuries of unstable climatic shifts morphed into much more significant cooling: a little Ice Age of sorts. Until 450, the North Atlantic Oscillation was in the positive mode, but in the late fifth century, the NAO index flipped to negative, which displaced long-established storm tracks to the south. Rainfall increased over much of the Mediterranean. At the same time, the volcanic quietude of previous centuries was shattered when strong eruptions made themselves felt. The year 536 was "a year without a summer," with little warmth from sunlight and a darkened sun caused by the high levels of volcanic ash in the atmosphere. In the eastern part of the empire, the cold, sunless year decimated wine harvests.[22]

The Italian statesman Cassiodorus observed a blue-colored sun.[23] Crops failed in Italy but distributions from the previous year's abundant harvest compensated for them. The year 536 brought famine to Ireland far to the north and unaccustomed summer cold as far away as China. By combining ice cores, tree rings, and physical evidence for global volcanic eruptions, we can now be certain that the 530s and 540s were decades of most unusual and punishing volcanic activity. The huge Northern Hemisphere eruption in 536 that brought ash-filled skies to Constantinople by March coincided with the coldest year in 2,000 years. Average summer temperatures in Europe dropped by as much as 2.5°C. An even more violent tropical eruption in 539–540 brought temperature drops of about 2.7°C to Europe once again. The cold was more severe than at the height of the Little Ice Age in the seventeenth century.

Fortunately, a good harvest in 535 CE provided some temporary relief from famine, as did the traditional resilience of Mediterranean farming societies to crop failures. The immediate effects were subtler than merely widespread hunger. The cooling of what is often called the Late Antique Little Ice Age—a clumsy label at best—increased the great stress felt by imperial authorities already under great pressure from plague attacks and intensive attacks in Europe from the steppes. A deep decline in solar activity had begun in about 500, bringing less heat from the sun to earth. The decline in solar output from the mid-530s to the 680s came just as the volcanic eruptions affected global temperatures. The plunge in solar energy was greater than that of the infamously frigid Maunder Minimum of the seventeenth century, detailed in Chapter 13.

As is always the case with climatic shifts, the effects differed from one region to another. The NAO flip had pushed storm tracks southward, which brought abundant rains and floods to Italy and Sicily. Intense snowfall, lower temperatures, and more rain affected a broad swath of Turkey (Anatolia) and further east. More frequent frosts caused olive trees to freeze and die in many areas where they had traditionally grown. North Africa witnessed disastrous aridification, with the great city of Lepcis Magna being abandoned, its buildings mantled in sand. No longer was North Africa a granary.

Justinian was a proactive emperor who put great energy into combating the droughts caused by climatic shifts such as persistent drought. He constructed aqueducts, cisterns large and small, and strategically distributed granaries. The emperor improved the transport of grain, reclaimed floodplains, and moved riverbeds. As one writer put it, he "joined forests and glens to each other" and "fastened the sea to the mountain." Justinian seems to have assumed that he could subdue the environment like one of his subjects. But the massive climatic shifts of his day were far too powerful for a mere mortal to conquer.

Justinian survived the twin catastrophes of environmental change and the plague, but the climatic extremes of the Late Antique Little Ice Age gradually drove the empire to a critical tipping point. The moment came in different ways throughout the connected world of the empire. In the final analysis, the Roman Empire died slowly from within, triggered by environmental causes.

In the eastern Mediterranean, the Nile valley had become a heavily engineered, humanly organized oasis, intended by their Roman masters to serve primarily as a Roman granary. The crops came from an intricate jigsaw of canals, dikes, pumps, and wheels that depended on vast reservoirs of brutally hard human labor. The Egyptians relied on monoculture, on wheat demanded by Rome and Constantinople to the exclusion of virtually everything else. When the bottom fell out of the wheat market as a result of the plague that left fewer people in Roman cities to feed, the glut of freshly harvested grain caused great economic suffering.

A sense of impending doom spread across the Roman Empire. The battering ram of apocalyptic events seemed to be a chronicle of God's anger and judgement, as he chastised the devout. From the sixth century, we have the earliest records of Christian penitential processions designed to expiate various communities' sins. Pope Gregory the Great of Rome organized a great rogation—three days of prayers and chanting. Choirs sang the psalms; prayer lines crossed the city. Eighty people are said to have died as the prayers continued uninterrupted. Such ceremonies were calls to repentance. But, in the end, a new version of Abrahamic monotheistic ideology from Arabia prevailed as the armies of Islam detached the eastern Roman possessions from the empire. Constantinople's lifeline of Egyptian grain ceased to operate. For many centuries, the empire had survived on a tightrope of fragility and resilience. But, finally, the inevitable forces of the natural world undermined the people of the empire, who could bear suffering no more.

By any standards, the Roman Empire was an enormous, complex enterprise with great wealth at its disposal. The emperors faced many challenges that confronted their profoundly traditional, quite well-integrated domains. The decline was gradual, beginning in the second century CE and enduring until the eighth. As the great eighteenth-century historian Edward Gibbon pointed out, it took longer for the empire to decline than it took many states to rise and fall.[24] The process of implosion was no sudden collapse but rather a slow-moving transformation from a tightly controlled, relatively centralized empire into a mosaic of different societies and political entities that either suffered, even vanished, or prospered. Rome flourished on the backs of commoners, especially slaves, and was influential over a huge area because of its military organization and its efficient infrastructure for moving food and other commodities over long distances. The empire was a catalyst for vulnerable exposure to both short- and long-term climatic change. Relatively brief climatic events that lasted a few years or a decade or two were manageable because of the efforts toward transportation and centralized storage of foodstuffs. Heightened vulnerability came with much longer arid cycles, especially megadroughts, which wreaked havoc on both local sources and imported grain supplies. Add to this the crowded, unsanitary conditions of Roman cities, large and small, and pandemics like the Antonine and Justinian plagues were inevitable, and they played a decisive role. But while both climatic events and plagues were tipping points, one should never forget that economic and social upheavals, as well as military events, were often the metaphorical ripples that radiated from the shock of unexpected climatic events.

All preindustrial civilizations depended on human labor and subsistence agriculture. Intensification of agriculture to feed growing urban markets and standing armies and the widespread use of food rations to feed workers, armies, and regiments of officials: these were among the developments that heightened the vulnerability of increasingly complex societies to climatic shifts. Food surpluses were always important to risk-averse subsistence farmers, who tilled the soil in constant fear of famine and malnutrition. In contrast, growing cities and empires relied increasingly on what was effectively monoculture in crops like wheat that were sensitive to drought, cold, and also excessive rainfall. Rome and Constantinople came to rely heavily on imported grain, grown in distant lands where monoculture of basic food crops became a near-industrial activity. The citizens of both cities and of other major population centers, as well as armies and bureaucracies, relied on government handouts, which ensured political and social tranquility. The Nile valley, parts of Europe, and North Africa became the Roman Empire's granaries. This was fine during well-watered decades, but things fell apart when the Egyptian inundation faltered and drought ravaged North African fields. Granaries emptied; hunger ensued; grain riots resulted. The growing chasm between the wealthy elite and often hungry commoners widened inexorably in the face of climate change and plague. A different, more fragmented world emerged from the disintegrating remains of the Roman Empire. The state gave way to more meaningful local, cultural structures that fashioned the world in new ways.

The Roman Empire expanded and expanded until it extended from northern Britain to Mesopotamia, with trading contacts far further afield. The expansion, which occurred for the most part during centuries of relatively favorable climatic conditions, involved incorporating multiple cultures and economies into a single, vast system. Many of the political actors are well known: Julius Caesar, Cleopatra, and emperors with varying strengths and weaknesses such as Augustus, Claudius, Nero, and Hadrian, among others. The empire flourished against a background of economic, military, and political strategies, which were remarkable in that those who initiated them devoted little time to thinking in the long term. They certainly took little account of long-term environmental changes that would unfold beyond their lifetimes. We often do exactly the same today, despite being able to discern potentially catastrophic climatic changes on the horizon.

The strategies of the later empire had to be reactive, for, unlike us, Rome had no warning of major climatic changes, including major droughts.

Looking back at the disintegration and transformation of the Roman Empire, it's easy to find startling parallels with today's globally focused world, where the problems confronting us are far greater. We have much to learn about the dangers of vulnerability to climate change from the emperors of nearly 2,000 years ago. One just has to look at the globalization of today's food chains to get the point. Compared with the Romans, we have the potential ability to adjust our food chains in the face of major climate changes. But there's always the possibility that future warming will be too rapid and on such a large scale that tens of thousands, even millions, of us will starve. And almost no one thinks about this in political terms.

6

THE MAYA TRANSFORMATION

(c. 1000 BCE to the Fifteenth Century CE)

The Romans were lucky. Their empire prospered and expanded to its greatest extent during the four centuries of relatively stable, warm, and wet conditions that spanned much of the Mediterranean world after 200 BCE. They created a far-flung realm based on intensive agriculture, largely unaware of the hazardous environmental foundations that supported their seemingly invincible edifice. The empire seemed destined for immortality, a dominant entity that would survive forever. Many believed that the demise of the empire, if ever it occurred, would mark the end of the world.

Devout Romans assumed that the future of humanity was in the hands of the divine, be it numerous deities or one. This was why, just like many ancient rulers, the Roman emperors stressed their close relationships with divinities. However, as we saw in the previous chapter, the gods failed to intervene in the harsh realities of climatic instability after the third century, which ultimately undermined an empire beset by complex climatic, political, and social pressures—plus disastrous pandemics. The great cities, Rome and Constantinople, survived, though much reduced, in a transformed medieval world, surrounded by expanding Islam. Slight changes in the tilt of the earth's axis as it circled the sun and powerful volcanic activity helped create the volatile and precarious European and Mediterranean world that led to the so-called Dark Ages. But before entering this maelstrom of climatic changes, politics, and warfare, we must travel further afield, for the warmer, stabler environmental conditions of the early first millennium CE also helped impressive civilizations in the Americas.

Both the powerful city-state of Teotihuacán in the highlands of central Mexico near Mexico City and the diverse Maya civilization in the lowland Yucatán achieved greatness in Mesoamerica during the first millennium CE.[1] Maya rulers claimed godly ancestry and governed using astute combinations of commercial acumen, expert diplomacy involving political alliances and marriages, and occasional warfare among elite warriors. They presided over volatile kingdoms that rose and fell with bewildering rapidity. At its height, Classic Maya civilization, which lasted from about 250 to 900 CE, encompassed about forty cities and kingdoms.[2] But during the tenth century, Classic Maya civilization in the southern lowlands, modern-day Guatemala's Petén, disintegrated. Royal dynasties collapsed, cities imploded, and their inhabitants dispersed into rural villages. Large numbers of people moved southward into what is now Honduras, much as Indus people moved into Rajasthan as their civilization dissolved. Once densely inhabited farmlands reverted to forest, with little subsequent recovery.

The transformation of Classic Maya civilization has fascinated scholars for generations, but only within the past twenty years or so has climate change, with its droughts and floods, become a major player in this scholarly debate. New generations of more accurate climatic data tell a complex and intricate story that goes far beyond droughts and hurricanes.

LOWLANDS AND LORDS (C. 1000 BCE TO C. 900 CE)

The central Maya lowlands of the Yucatán Peninsula are a challenging environment for subsistence farmers based on dispersed farmsteads in scattered communities, let alone complex, highly competitive city-states ruled by ambitious lords.[3] Yet the Maya farmed and survived on this once densely forested plateau of porous rock that forms the spine of the peninsula for over 2,000 years. The realities were daunting. Seasonal rainfall is highly unpredictable, falling, usually in short-lived storms, during the hot summer months. Winters are dry. Water drains quickly through the bedrock. Almost all the lowlands lack any form of permanent water supplies except for widely separated springs. Such

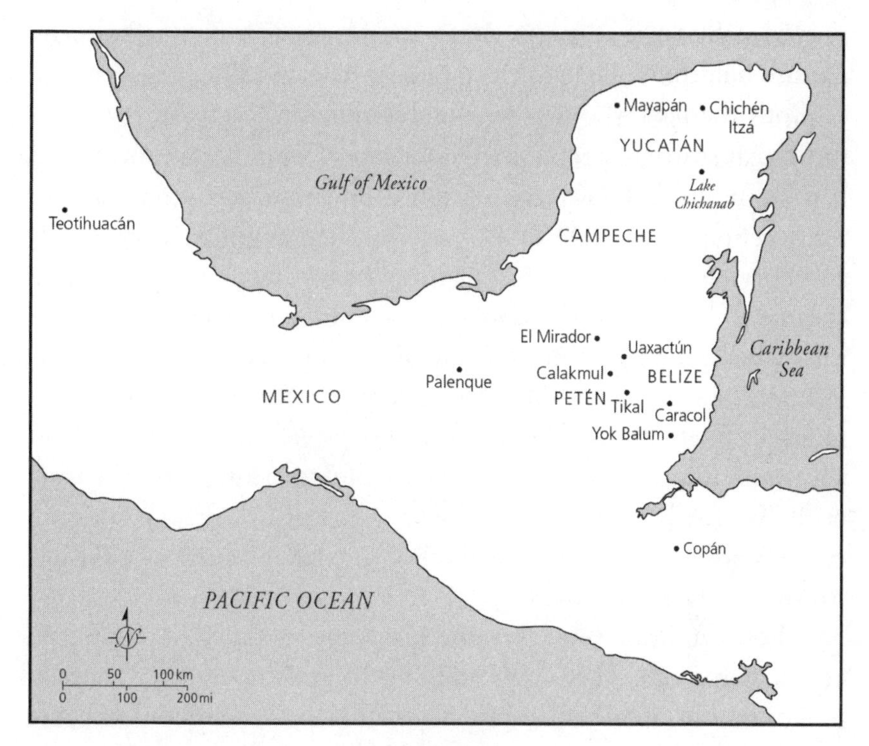

Archeological sites mentioned in Chapter 6.

aquifers lie one hundred meters or more below the surface, making them inaccessible. Add sometimes decade- or century-long droughts to the equation, and water becomes the most critical element in the survival equation. At these times evapotranspiration—the movement of water from the sea, lakes, plant canopies, and other sources, which exceeded rainfall—was all-important.

Dense seasonal rainforest mantled the landscape, where it had not been cleared for cultivation. The growth flourished in fertile soils of varying depth. Depressions filled with dense clays up to a meter thick filled with rainfall runoff during the wet months, creating valuable seasonal wetlands. Phosphorous is the limiting nutrient for vegetation, captured by the forest canopy and washed into the soil. Producing the increasingly large food surpluses required by proliferating city-states and their demanding leaders required diverse, efficient farming—and an intimate knowledge of the complex lowland environment.

Between 1000 and 400 BCE, large numbers of Maya farmers moved onto the Yucatán lowlands, many from the Gulf Coast, where Olmec societies thrived. Indigenous farming had long flourished in the Yucatán, where people were cultivating domesticated plants and had an intimate knowledge of forest environments acquired over many centuries.[4] By 600 BCE, they were building massive pyramids and burying ancestors within platforms and other structures. These became sacred places where they venerated their forebears, and genealogy became important as a way of claiming ownership of places. Within a few centuries their descendants constructed massive complexes of finely finished buildings with plastered masks of gods and ancestors. Thus was born the institution of divine kingship, *ch'ul ahau*, or "holy lords," as illustrated by the great city of El Mirador. For generations, local farmers there had relied on wetland gardens built between 150 BCE and 50 CE.

These centuries were when the Maya first modified the landscape on a massive scale. They now had to feed a growing number not only of farmers but of non–food producers. They moved millions of cubic meters of soil to create reservoirs, canals, and ponds to store water for the dry season. El Mirador, which covered sixteen square kilometers in its prime, lies in a depression and depended on water supplies from it. As the population rose, so did the need for communal labor for both landscape modification to feed people and the creation of public buildings. Over the generations, social inequality became the norm, with a privileged elite, often related to the ruling lord, distancing themselves from commoners.

El Mirador fell apart suddenly, partly because of excessive deforestation and partly due to runoff and erosion that destroyed the surrounding wetlands, victims of plentiful rainfall. For centuries, local farmers had relied on wetland gardens to grow large grain crops, the food surpluses the city needed. With the city-state's large population of nonfarmers, its political and social infrastructure was threatened when the commoners could not feed the elites. For everyone, the only strategy was mobility—moving out into smaller settlements in the countryside as the urban center withered. By 250 CE, the center of Maya political gravity had shifted to the central lowlands, where new centers such as Calakmul and Tikal developed into powerful city-states in a time of more abundant rainfall. We know of their lords from deciphered glyphs, which reveal an ever-changing mosaic of diplomacy, trade, and warfare. Everything revolved around the institution of kingship, which passed from father to son, or from brother to brother, in dynastic lines that led back to a founding ancestor. Maya civilization was never a highly centralized state like Egypt or the Roman Empire; rather it was a montage of political units, large and small, that eventually morphed into four major city-states and numerous smaller kingdoms. This was a highly competitive society, dominated by powerful dynasties based at major centers like Tikal, Calakmul, Palenque, and Copán.

Tikal and nearby Uaxactún stepped into the political vacuum left by the decline of El Mirador. During the first century, an elite came to power at Tikal, where hieroglyphic texts reveal thirty-one rulers from 292 to 869 CE—some 577 years of dynastic rule. The newly powerful city-state became a multicenter kingdom before being conquered in 557 by the lord of a rising state, Caracol, now in Belize.

By 650 CE, the major lordly dynasties presided over elaborate public ceremonies that validated their spiritual ancestry and political power. They linked their actions to those of the gods and the ancestors, sometimes legitimizing their descent by claiming it reenacted mythic events. They connected their history to the present and to the supernatural otherworld, with society embedded in a matrix of sacred place and time. A Maya lord took care to proclaim that he was an intermediary between the living, the ancestors, and the supernatural world. This was the foundation of an unspoken social contract between the rulers and the ruled, the thousands of Maya who supported a tiny elite at huge environmental cost. Population densities rose dramatically over the Maya lowlands. Moderately fertile rainforest soils produced ever-lower yields. Even short drought cycles endangered precious water supplies, despite long-established practices like crop diversification. It was a matter of time before the land could not support a huge population of nonfarmers.

Not that the lords were unaware of the dangers of climatic shifts. Quite the contrary, for they ruled through centuries of gradual drying. Much of their ceremonial life revolved around water and rainfall. Ingeniously, the rulers of Tikal built temple pyramids whose sides channeled rainwater into reservoirs to water nearby fields and to control excess flow throughout the year. Maya rulers responded to population growth by building reservoirs and often quite extensive water-control systems to store water against dry years.

MAYA FARMERS, THEN AND NOW

Between the third and tenth centuries, hundreds of Maya settlements large and small proliferated across the lowlands, supported by remarkably diverse forms of agriculture. These included slash-and-burn farming (*milpa*) on cleared forest soils, terrace cultivation on slopes, and use of raised fields based on swamps and wetlands—farming remarkable for its aggressive manipulation of the environment and its scarce water supplies. Many farmers also had diverse house gardens with a plethora of plants and trees. At the local level, Maya farmers managed forests, stored water, and took full advantage of the varied soils and food resources throughout the lowlands. They were so successful that they managed the risks of a difficult environment for 4,000 years. They did so by knowing the most intimate details of their landscapes and by building and sustaining intensive food production systems that withstood at least two prolonged droughts before their civilization ran into serious problems during the ninth century.

Fortunately, the descendants of the Maya farmers of lowland Mesoamerica still flourish in their demanding, lowland environment. Many of the practices used by modern-day villagers persist from much earlier times, which means that they can provide us with insights into how people coped with droughts, crop failures, and other unexpected climatic woes. The diversity of modern Maya farming is striking, a response to everything from rising population densities to the qualities of local soils and changing rainfall patterns. Even the mixture of crops changes from year to year and season to season, depending on environmental conditions. For example, the Kekchi Maya of Belize still rely on traditional agriculture. They cultivate raised fields in poorly drained areas, terrace hillsides, and use the *milpa* system of slash-and-burn farming during wet seasons.[5] Kekchi dry season riverbank farming is an example of opportunistic ingenuity that requires long experience. Each farmer must balance the climatic conditions and regrowth of vegetation with other tasks. The earlier the corn is in the ground the better, for it must have a good start while the soil is still damp. The onset of the dry months varies considerably, which complicates matters, as does the critical task of harvesting the *milpa*. If the *milpa* harvest is good, then there is less time to plant in the dry season. A poor return means more time for clearing and planting.

Riverbank farming like this operates within a much larger-scale subsistence cycle. The key word is "cycle," for this helps us to unlock the strategies used by the Maya and, for that matter, many other subsistence farmers as they deal with irregular climatic variations year after year. Such a cyclical existence means that people living close to the land think of time as an endless circle. Their ancestors experienced the same rounds of planting, growth, and harvest, then a quiet season. Such a life had an unchanging permanence that depended on crops and rainfall.

This powerful assumption gave a central role to revered forebears. There were compelling reasons why Maya lords stressed their close relationships to divine ancestors and why Egyptian pharaohs conducted elaborate public ceremonies that validated their roles as divine rulers. Lordly relationships with ancestors tended to be magisterial and obsessed with spiritual legitimacy. The ties between ancestors and the living permeated deep into village life, where survival depended—and still depends—on a very close relationship with the environment, with rainfall, vegetation, and soil fertility. Today, the Kekchi rely on a mixture of common sense, intimate environmental knowledge, and a deeply held belief that the ancestors' experience is a priceless legacy for survival.

And in this region, what a legacy. In former times, this area had been densely occupied, highly dependent on the skill of its agriculturalists—and on the rains.[6] Populations peaked between 700 and 800 BCE. At that time, concentrations of 600 to 1,200 people per square kilometer were not unknown. A staggering eleven million people are estimated to have inhabited the lowlands. Most lived not in the grand cities themselves but throughout nonurban areas in individual household compounds distributed widely across the landscape. This pattern is not dissimilar to what was happening around the city of Angkor in Cambodia, which we shall visit in Chapter 9. Alas, both in Cambodia and here, any environmental stress in the hinterlands increased the chance of major social and political upheaval. By the eighth century, ancient Maya civilization in the southern lowlands was about to decline.

If you had lived in the Maya lowlands during the late eighth century, you would have dwelt in a human-modified, gradually drying landscape radically different from that of centuries earlier. The cumulative effects of environmental modification had accelerated as populations rose and crop yields declined. A mosaic of cleared areas, managed forests, fields, and cities had turned most of the lowlands into an engineered landscape. Of course, dense populations almost always result in deforestation, and fewer trees lead to higher temperatures and less capture of rainfall. Furthermore, the burning of the wood and crops results in more ash and pollutants in the air.

As settlements rose across the lowlands, the areas of impervious ground surface expanded dramatically. Both building activity and the spread of cultivated land further decreased the capture of phosphorous and increased sedimentation. In earlier centuries, upland sediments cascaded into river floodplains where wetland farming was productive, but the farmers reduced sediment loss by widespread use of terraced slopes for farming. Just maintaining field systems, as well as the proliferating canals, ponds, and reservoirs, required thousands of commoners and the labor forces of entire villages. So did the routine tasks of manuring, mulching, and weeding.

The long-term effects of deforestation alone were devastating. By 600 BCE, much of the forest in northern Guatemala's Petén had been felled. Forest clearance continued until most tree cover had vanished by the ninth century CE across the human-modified landscape. The long-term effects of a combination of continuous deforestation, land-use changes, and environmental degradation caused by Maya farming resulted in less rainfall, higher temperatures, and increasing water shortages.[7] These were quite apart from natural drought cycles. But once the deforestation became near total at a time of severe droughts, the continuous adaptive strategies used by the farmers were unsuccessful. Political instability and social disorder ensued, and Maya civilization disintegrated. A tipping point in the human and environmental system was reached, leading to cultural decline and eventual depopulation.

BEYOND THE TIPPING POINT (EIGHTH TO TENTH CENTURIES CE)

The decline of Classic Maya civilization in the lowlands came about through the concatenation of stresses in the changing relationship between people and the environment in conjunction with a spike in drought cycles. But far more was involved than merely such basics as food supplies and water. The moment had arrived when the parameters of maintaining Maya sustainability were just too complex and overwhelming—for rulers and elites anyway. Embedded as the elite were in the complex socioeconomic, ideological, and political dimensions of Maya society, the obstacles to maintaining or growing the system were enormous—and perhaps so large that doing nothing was easier. There was no single cause for the major ninth-century transformation of Maya in the central lowlands, determination of which remains a major controversy.[8]

Climate change, specifically drought, has long been a major and tempting villain ever since cores from Lake Chichancanab in the northern lowlands showed that there were significant droughts between 800 and 1000 CE.[9] The lake cores revealed that drier conditions prevailed between 750 and 1100, with multiyear droughts in 760, 810, 860, and 910, identified from a deep-sea core in the Caribbean's Carioco basin. The lake and ocean cores were not, however, as accurate as necessary. As a result, many experts tended to downplay climate's role in the decline of Classic Maya civilization.

A new generation of research has produced much more accurate information on droughts and rainfall. In the southern Maya lowlands, a fifty-six-centimeter-long stalagmite from Yok Balum Cave has provided an accurate 2,000-year climatic sequence.[10] The Yok Balum aragonite (a calcium carbonate mineral) stalagmite is especially important because it grew quite fast and continuously for two millennia. The researchers obtained no fewer than forty uranium series dates for rainfall, accurate to within five to ten years, that agree well with climatic data from other sources. Between 440 and 660 CE, the region received unusually high rainfall, followed by three and a half centuries of gradual drying. This culminated in a lengthy and very severe drought between 1000 and 1100, the worst in 2,000 years. That was not all. A major drought lasted for a half century between 820 and 870, with another minor dry interval in about 930. The climatic information from Yok Balum corresponds well with records from elsewhere in the lowlands of a severe drought between 820 and 900 and another between 1000 and 1100.

By any standards, these droughts, known from a range of evidence, were periods of prolonged aridity, which must have had a serious effect on farming communities in a region of unpredictable precipitation. Dry years have an obvious effect on crop yields and agricultural productivity. These consequences are especially serious if the wet season arrives late or if there is total crop failure. The droughts of the late first millennium were another matter. These persisted for decades, even centuries.

As archeologist Doug Kennett and climatologist David Hodell point out, there's an important distinction between agricultural and hydrological drought. The former results from lack of rain and increased evaporation that dries out the soil, leading to crop failure. Meanwhile, lake levels, stream flows, and groundwater supplies may not be affected for some years. The Maya were well aware of the need to conserve water supplies. Such strategies were effective in the short term and for slightly longer periods, but much depended on the density of population consuming the water sources. When dry cycles persisted or were unusually severe, hydrological droughts kicked in as water supplies dried up or became scarce. There could be serious socioeconomic consequences, especially when rising population densities and demands for water and other resources in the environment came to exceed the supply.

Far more than merely drought was involved. Maya society was a social pyramid, ruled by a tiny elite who enjoyed power through a combination of force and well-orchestrated ideology. They enjoyed a much higher standard of living than artisans and commoners, with virtually all the wealth concentrated in lordly hands, as were control of vital resources, like obsidian and salt, and sophisticated knowledge such as astronomy, mathematics, and calendrics. The unspoken social contract with the people was a guarantee of ideological, material, and spiritual authority. Problem solving became increasingly fraught as the elaborate mechanisms of governance became ever more elaborate and conservative.

Maintaining and legitimizing the authority, political power, and wealth of the elite was an ever more complex task, covering everything from maintaining infrastructure to reclaiming wetlands and administering military forces for defense and raiding neighbors. These were monarchies, ruled by ideologically rigid, powerful lords, thought of as having semidivine powers. Apart from their own lavish establishments, they demanded large food surpluses to maintain their courts, a hierarchy of functionaries, and an entrenched elite in the style to which they were accustomed. Military campaigns required support. So did numerous skilled architects, artisans, and scribes, nonfarmers whose rations in food and other commodities supported their work. A major staple crop was maize, a food so important that it played a prominent role in public and private rituals as well as art. But maize is a tropical crop that is nearly impossible to store in a humid environment like that of the lowlands. Other crops included beans, squash, and chili peppers, but whatever the food, every Maya farmer had to feed his or her family and save enough seed for the next season. Furthermore, every farming family provided both food and labor for the ruler and the elite to maintain the increasingly demanding and intricate superstructure of a jigsaw of competitive kingdoms. Add the varied mosaic of different crops and productive soils, topography, and, above all, water supplies, and the task of administering a response to even short-term climate change rapidly became overwhelming.

The end of the eighth century saw rulers unable to deliver on their social promises, especially providing clean water via massive reservoirs as drought continued. By then, the centuries-old economic and political structure with its semidivine lords was in serious decline. The demands of the rulers on the ruled created an immediate and continuous tension between the haves and the have-nots in a society riven by intense competition and factionalism. Everything depended on what was ultimately a potentially unsustainable subsistence agricultural society living in a region plagued by inadequate rainfall and unpredictable, prolonged droughts.

The political fallout from powerless authority was enormous. For all its diversity, ancient Maya society shared many cultural traditions, including the all-important institution of divine kingship. Such kings or queens were the major actors in the volatile relationships between the larger kingdoms and numerous hierarchies of smaller domains with ever-changing loyalties. Every Maya ruler lived in a politically charged landscape of transitory alliances and trade networks, as well as ties of ancestral kin. But, in the final analysis, loyalties and cultural ties were local, which also made sweeping initiatives to mediate climate change almost impossible.

COPÁN DISSOLVES (435 TO 1150 CE)

As powerful city-states dissolved, artisans and commoners dispersed into the hinterlands or moved elsewhere in search of opportunity. For example, Copán in Honduras is a spectacular Maya center, adorned with pyramids and plazas covering twelve hectares.[11] For four centuries after December 11, 435, a powerful dynasty ruled the kingdom, founded by lord K'inich Yak Ku'k Mo' (or "Great Sun Quetzal Macaw"—the macaw and quetzal being brightly feathered birds).

Long-term field surveys around Copán have chronicled dramatic population shifts during the four hundred years of the "sunbirds" dynasty. Between 550 and 700 CE, there was a rapid population boom. People lived close to the central precincts and their immediate surroundings, with only a small rural population. Both the population and social complexity increased until 18,000 to 20,000 people lived in the Copán valley, with about 500 people per square kilometer in the central core. The population seems to have doubled every eighty to one hundred years. Rural populations were still scattered, but farmers were now cultivating less desirable foothill soils to stimulate crop yields.

But change was afoot. In 749 CE, a lord named Smoke Shell ascended the throne of what had once been a great city. He embarked on a frenzy of building at a time of intense factionalism and internal tension, some of it triggered by the realities of declining rainfall. The political order seems to have changed, as lesser nobles commissioned inscriptions for their houses as if they were asserting themselves in a time of declining political authority. Profound demographic and political changes ensued. The Smoke Shell dynasty ended in 810, just as urban depopulation began. About half the urban core and periphery's populations were gone within four decades, but the rural population increased by 20 percent. Small regional settlements replaced the great center, as the cumulative effects of overexploitation of even marginal agricultural soils and uncontrolled soil erosion came home to roost. In 1150, no more than 5,000 to 8,000 people lived in the Copán valley.

Copán's dispersal was a logical response to falling crop yields and the rapid growth of urban living, also a traditional reaction to severe drought, as it was in many ancient societies. The dispersal was not unique. Long-term surveys in the hinterlands of major centers like Tikal and Calakmul have provided ample proof of the decline of densely concentrated urban populations. After the eighth century, huge areas of the southern lowlands fell into disuse and were never reoccupied, even after the Spanish *entrada*. The expanding Maya population had depended on an agricultural system that made no allowance for long-term problems such as prolonged droughts. At the civilization's height, perhaps eleven million Maya dwelt in the lowlands, more people than live there today. By then, the system could no longer expand or produce the kinds of riches that the rapacious elite demanded. Once influential city-states could only decline and disperse, just as was the case at Copán and Tikal.

Much of the literature on the Maya dispersal gives the impression that the disintegration was universal. This was most emphatically not the case. Some city-states survived on a reduced scale. Others continued to flourish, especially around important rivers and astride major trade routes. Coastal centers survived, especially in the northern Yucatán. Powerful economic and social factors were in play, among them access to coastal and river trade routes, the prevalence of warfare, and, perhaps most important of all, a major shift in trading activity from inland to maritime.

Droughts and crop failures intensified competition for food supplies and control of trade routes. Vicious warfare flared in many places during the seventh and eighth centuries, but not necessarily as a result of aridity. Maya lords depended on maize crops to maintain their power. Crop yields increased until the temperature reached about 30°C in dry cycles. Thereafter, yields declined rapidly, as did reservoir levels. As the number of days over 30°C increased, food supplies plummeted, threatening royal power. In response, ambitious rulers attacked other kingdoms in the belief that successful campaigns would restore their seemingly declining legitimacy. Dry cycles may also have reduced violence because both food and water were in short supply, making provisioning for warfare much harder. But whatever the temperature conditions, violence was endemic at times during Maya history, so much so that some nobles fleeing violence built defensive walls around large tracts of agricultural land to protect their growing crops rather than fortifying temples and other imposing structures constructed earlier.

DISINTEGRATION (EIGHTH CENTURY ONWARD)

Warfare may indeed have played a role in the disintegration of southern lowland Maya society, but there is no question that drought played a significant role in destabilizing it. Historical dry cycles recorded in the Yok Balum speleothems coincided with crop failures, hunger, and starvation, as well as with outbreaks of famine-related diseases. There is also evidence of population decline and of people moving into smaller settlements. This was a classic mobility strategy, significantly used once again at a time when the aridity was far more prolonged and severe than those of earlier times.

What actually happened? The gradual disintegration of Classic Maya kingship was not a dramatic event. Rather, between 780 and 800 CE, long-established political and social networks in the southern lowlands began to fall apart, while warfare intensified.[12] The result was what Doug Kennett and his colleagues call "balkanization," as political networks became decentralized and populations dispersed. It was not so much a collapse as a reorganization of society, reflected in the survival of writing, the calendar, and other valued cultural traditions long after 900 and until the Spanish *entrada*.

The most dramatic transformation occurred among Maya kingdoms centered in northern Guatemala, western Belize, the southern

Yucatán, and the Copán region of Honduras. They left behind an agrarian landscape that is still barely inhabited forest. In the central lowlands, the forest recovered, but the people never returned, to the point that the rainforest became a refuge for Maya fugitives from Spanish rule. Even today, the population density is one or two times less than that during Classic times. Quite why remains a mystery. Widespread forest clearance ceased until the hardwood logging of modern times. Small parties may have ventured into the overgrown landscape to exploit economically valuable trees, such as ramón trees, with their nutritious fruits and nuts, a valuable food source in rainforest environments prone to drought. Perhaps the human costs of clearing the forest and restoring the infrastructure of intensive agriculture were too high.

EVENTS IN THE NORTH (EIGHTH CENTURY ONWARD)

Maya civilization continued to flourish in the northern Yucatán.[13] A powerful kingdom based in Chichén Itzá prospered between the eighth and eleventh centuries, due in part to new subjects escaping the increasingly dry southern interior. This rise to power is hard to believe when one realizes that surface water supplies were much rarer in the north. Chichén Itzá's power came from aggressive expansion and alliance building, as well as from its control of seaborne trade and wide contacts across the Maya world. In this case, the response to aridity was predominantly economic and political and so effective that Maya civilization revived—in a different manner though, focusing on shared rulership.

Chichén Itzá's dominance faltered during the eleventh century in the face of the longest and severest drought in the region, which undermined a long-established status quo. But around 1220 CE a new state arose based at Mayapán, which lay in the interior of the north.[14] Mayapán, with some 15,000 inhabitants, was the major political capital of a powerful regional confederacy. This was a cosmopolitan resurgence of Maya civilization, marked by spectacular architecture, widespread external contacts, and a revival of traditional religious beliefs commemorated with magnificent codices. Located as it was near a ring of *cenotes* (natural sinkholes), a plentiful underground water source,

Mayapán flourished until about 1448, contending in later times with a century and a half of severe drought episodes. These played havoc with food supplies, disrupted market networks, and led to political instability accompanied by warfare.

But Maya civilization survived, partly because the major centers were not closely interconnected, which made them less vulnerable to the kinds of political turbulence that had upended the south. Right up to the Spanish *entrada*, impressive coastal towns prospered, and sophisticated market systems operated over large areas. Everything resulted from successful adaptations to local environmental challenges, regional droughts, and food shortages. Society changed constantly in a literate Maya world with centuries of cultural tradition behind it. The arrival of Spanish conquistadores in the early sixteenth century changed the trajectory of Maya history as the people adapted to new economic, political, and spiritual circumstances.

The so-called Classic Maya Collapse is a misnomer, dramatic as it may sound. Rather, the civilizational decline was a complex process of faltering adaptations to lengthy drought cycles that took many generations. That said, the Classic Maya political system did collapse, while farmers moved on. In the end, ancient Maya civilization had undergone a transformation that followed what appears to have been a critical social, political, and ecological tipping point around 800 CE. The interactions between the Maya and their landscape produced various degrees of environmental stress that coincided with severe drought. There came a point when Maya rulers, for all their elaborate ideology and tight control of society, were unable to organize adaptations to much drier lowlands. The daunting task of organizing bold steps to adapt to a subsistence crisis in city-states riven by factionalism and warfare overwhelmed great lords in all their arrogant magnificence. The people lost faith in their authority and in the social contract between the rulers and the ruled, which fell apart. So they dispersed.

On a global scale, with millions of people involved, in a world riven by petty nationalisms, we face a threat of anthropogenic global warming and potentially catastrophic climate change unimaginably greater than that faced by the Maya lords. Their subjects moved away to rural farmsteads or sought opportunities elsewhere, for the impact of the crisis varied from one region to another. But the lesson of the Maya experience is crystal clear: strong, decisive leadership matters. Many people are working to confront the issue of future climate change, but we have lacked the powerful, forward-looking leadership that transcends generations. We are in real danger of a fate like that of the rulers of Tikal and other great Maya cities, for not only do many of us deny the impending climate crisis, but most people feel overwhelmed by the challenge as we approach a similar environmental tipping point but on a vastly greater scale. The Maya experience reminds us that a great deal of climatic adaptation is local and that inaction in the face of climate change is not a viable strategy.

This is why *local* measures to cope with climate change are far more effective than grandiose administrative schemes developed by anonymous officials concerned wholly with crop yields. What matters far more is risk management, above all at the local level, something we often ignore today.

7

GODS AND EL NIÑOS

(c. 3000 BCE to the Fifteenth Century CE)

Under blue skies, the white snow extends as far as the eye can see. We are at the remote Quelccaya ice cap, high in the Andes mountains of northern Peru, one of the largest tropical ice fields in the world. Today the ice cap covers about forty-three square kilometers, its highest point lying at 5,680 meters above sea level. Yet, at the close of the last Ice Age 18,000 years ago, it was far larger: inexorable human-caused global warming is shrinking the ice cap to the point that it may vanish by 2050. East of the ice cap, the mountains drop off into the Amazon basin, the tropical rainforest being only forty kilometers away. Unusually for mountain glaciers, the ice lies on a flat surface, in places two hundred meters thick. This makes it an ideal location for drilling ice cores, which have revealed well-defined layers—each representing a year—separated by dry season dust layers sufficient to reconstruct about 1,800 years of Quelccaya's climatic history.

In 1983, paleoclimatologist Lonnie Thompson from Ohio State University drilled two long ice cores in the central part of the field, using a solar-powered ice drill—other power sources were unavailable.[1] Since he didn't have the means to carry away the frozen cores, he cut them into samples, which he melted and bottled in the field—thus recovering sections up to 1,500 years old. In 2003, logistics had improved sufficiently for Thompson to transport two cores that went all the way to the bedrock, still frozen, to his Ohio lab. Thompson now has a Quelccaya climate history that goes back 1,800 years and reveals how both the El Niño Southern Oscillation (ENSO) and the position of the Intertropical Convergence Zone have influenced the climate of the ice cap.

El Niños bring westerly winds, which reduce the moisture that reaches the ice and carry heavy rains to coastal deserts along the west coast. The warming El Niño and its cooling sister, La Niña, interchange very irregularly over time. The former causes droughts to settle over southern Peru and Bolivia's high-altitude grasslands, or altiplano (Spanish for "high plain"). In contrast, La Niña conditions bring rainfall to the highlands. Together, they are powerful drivers of climate along the Andes and along the western coast of South America, particularly along the arid coastal plains of Peru. There, the runoff water from the nearby Andes enriched Peru's gold-rich preindustrial states, such as that of the Moche. ENSOs are complex events, which played a significant role in ancient Andean history.

THE COAST: CARAL, MOCHE, WARI, AND SICÁN (3000 BCE TO 1375 CE)

Two major poles of Andean civilization developed over many centuries. One great pole of ancient Andean civilization lay in the highlands, centered on Lake Titicaca. The other thrived far to the northwest, down on the lowland coastal plain of northern Peru, one of the driest places on earth. As a whole, this vast region comprises a series of west-to-east environmental zones—coastal desert and river valleys, mountains, highlands, plains, and tropical rainforest, among others. Each provided crops under different conditions, meaning that self-sufficiency and long-distance trade were enduring realities.[2]

The people of the coast depended heavily on the inshore anchovy fishery, which yielded both food and fish meal, much of which was traded to the highlands. Fishing was a vital task of lowland civilization. So was irrigation agriculture in river valleys. Virtually all of north coast Peru's irrigation water comes from mountain runoff that flows down rivers dissecting the coastal plains. The coast is a fragile environment, where catastrophic earthquakes occur, to say nothing of major, often lengthy droughts, desertification and sand dune formation, and strong El Niños that bring major floods. Living with these

Archeological sites in Chapter 7.

environmental hardships placed considerable restraints on coastal societies, except when changes, such as gradual desertification, permitted them to slowly adapt over long periods.

By 3000 BCE, some long-occupied settlements near the Pacific housed between 1,000 and 3,000 farmers and fisherfolk. They were close-knit communities with strong kin ties and a profound reverence for their ancestors, reflected in flamboyant, elaborately decorated textiles that depicted anthropomorphic figures, crabs, snakes, and other creatures. There were cities too, notably the ancient site of Caral (c. 3000–1800 BCE) in the Supe valley of Peru's north-central coastal region.[3] With its huge mud pyramids, plazas, residential houses, and temple complex, Caral was a formidable contemporary of the Old

World civilizations of the Indus valley, Egypt, and Mesopotamia. Its people shared Egypt's love of pyramids; yet, as with the Indus valley, archeologists have found no trace of warfare at Caral—no mutilated bodies, no battlements, no weapons. Instead, it seems that Caral was a peaceful city, a thriving metropolis that stretched over 150 hectares and spawned at least nineteen contemporary satellite sites. Precisely why the well-populated and well-connected Caral faded remains unknown, but throughout this region—as with every region of the world—cultures came and went, elements remained and elements disappeared, as people navigated social, political, and climatic changes. This interplay is beautifully illustrated when we travel forward in time to focus on the events surrounding some of the first millennium CE.

Around the time that Emperor Tiberius was throwing his enemies into the Tiber and Vesuvius was erupting, a rich, new culture emerged on the northern coast of Peru: the Moche state (c. 100–800 CE), ruled by a rich elite who buried their dead in mudbrick pyramids and left a legacy of golden treasures and rich artwork. They presided over a strip of coastline some four hundred kilometers long and only up to fifty kilometers wide, stretching from the Lambayeque valley in the north to the Nepeña valley in the south.[4] Of course, given the great cultural legacy of Peru, the Moche did not emerge from a vacuum. Rather, they built their state on a mosaic patchwork of well-established local valley irrigation systems. Their sites were laced with canals and irrigation systems, but everything depended on flexible farming methods based on individual villages. The Moche's agriculture base required small-scale labor and simple irrigation works that were, above all, easily repaired. Just as had been the case in the Old World, dispersed local communities relied on runoff from springs and occasional rainstorms.

The widely spread irrigation systems gave the state a measure of protection against prolonged droughts and heavy El Niño rains that could inundate and destroy entire irrigation systems in a few hours. The spring runoff from the Andes seemed like an annual gift from the supernatural world. Judging by Moche artwork and burials (they left no written texts), formidable and all-powerful lords ruled the state.[5] They claimed supernatural powers and acted as intermediaries between the people and the gods that nourished the coastal fisheries and precious crops. The Moche rulers, bedecked in all their gold and silver finery, appeared at elaborate public ceremonies to reinforce the belief that each leader was essential for life to continue. Without him the sun might not rise and the fish might die. As with (the slightly later) people of highland Tiwanaku (see below), the Moche subjects paid taxes to these "life-giving" leaders via the goods and foods they produced, as well as through forced labor, with large numbers of commoners deployed to build enormous monumental platforms and temples.

To us, the system might look like an elite-serving invention, a fairytale, a fakery, but the Moche took these ideas as seriously as life and death. In a world of uncertainty, before the rise of modern science, the leaders and their gods stepped in. The Quelccaya ice cores provide us with evidence of the harshness of coastal life, including a litany of severe droughts that reduced rainfall to as much as 30 percent below average.[6]

The most severe drought, from 563 to 594 CE, came at a time when the Moche rulers (or lords, or warrior-priests, as they are variously called by archeologists) lived downstream, close to the Pacific. This strategic location gave them control of both water and the rich anchovy fisheries close inshore, a lucrative source of nitrogen-rich fish meal traded by llama caravan to the highlands. Droughts turned irrigation systems into barren dust bowls. The lords used carefully husbanded grain stored by the state against dry years, but malnutrition must have been widespread. Fortunately, they could also rely on the fisheries until powerful El Niños struck at the height of the dry cycle. Warmer waters from the north decimated anchovy stocks, as torrential rains caused the desert rivers to become raging torrents, carrying everything before them. ENSOs devastated the Moche landscape; dozens of villages vanished under mud; adobe houses collapsed and their occupants drowned.

The warrior-priests were well aware of the effects of powerful El Niños. They responded by deploying people to rebuild irrigation systems and by performing human sacrifices. While investigating a secluded plaza by the Huaca de la Luna (Pyramid of the Moon) in the Moche valley, archeologist Steve Bourget uncovered dizzying wall depictions of seabirds and marine creatures—all associated with the warm ENSO current close offshore—but in the midst of this artistic explosion, he also found the skeletal remains of about seventy slaughtered warriors. He believes that the Moche rulers used human sacrifices and elaborate rituals as ways of reinforcing their authority in the face of disaster. Then another strong El Niño descended on the valley. Huge dunes formed by river sediments washed ashore, covered hundreds of hectares of farmland, and inundated the Moche capital. Both the Moche valley lords and their contemporaries in the Lambayeque valley moved upstream.

Despite these climatic reverses, the Moche still maintained extensive field systems constructed with as little investment as possible. The population became much more mobile, with smaller settlements built in different environmental settings instead of the large urban centers of earlier times. The farmers repaired damage promptly, as competition intensified for access to both fertile land and water.

Between 500 and 600 CE, the Moche consolidated their increasingly smaller, more dispersed settlements in the Andes foothills at the necks of coastal rivers, where they flowed into the desert.[7] By this time, Moche domains were increasingly fragmented, and any form of regional control of food production was difficult. A weakened leadership wrestled with sudden climatic shifts, as well as attacks from the highlands, as another severe El Niño washed away key field systems. The lords lost their divine credibility, and the Moche state came apart. Like Egypt's pharaohs, they had managed to survive at least one disastrous climatic episode that threatened their kingdom. But, unlike with the pharaohs, the environment allowed them little flexibility. The artificial landscape they created in the river valleys required long-term planning, technological ingenuity, and the abandonment of a rigid ideology that no longer held together a tightly controlled society. Apparently out of touch with life in the villages they governed, they ran out of options, and their rich, gold-laden society gradually dissolved into numerous smaller kingdoms after 650.

Among the patchwork of kingdoms was that of the Wari. Their domains extended from the Andean highlands down onto the northern (and possibly also central) Peruvian coast from around 500 to 1000 CE. A sophisticated culture, the Wari buried their elite with fine jewelry, as well as exquisite textiles and pottery. They worked the land wisely and developed a formidable system of terraced agriculture on hillsides. Yet, weakened by drought, they too ultimately fell. Interpersonal violence perhaps hastened their end: the blocked-up doors of some of their governmental buildings excavated at the city of Wari whisper of fleeing. Perhaps, suggest the archeologists, the citizens planned to return once the rains or the peace returned, but they never did.

Then there was the coastal Sicán culture. Its leaders came to power as Moche society fragmented in about 800 CE. Probably descendants of the Moche elite, they invested heavily in richly adorned ceremonial centers, dominated by artificial mountains fashioned of adobe bricks. A twenty-seven-meter-high pyramid, now known as the Huaca Loro, overlooked a large plaza and the Sicán center at Batán Grande in the Lambayeque valley. The richly decorated elite buried in shafts wore spectacular golden masks and ornaments. Commoners lay in shallow graves with little or no adornment. Like their Moche predecessors, they were just as vulnerable to the ravages of ENSOs. Batán Grande fell in the face of a massive El Niño just before yet another kingdom, Chimor, conquered the Sicán in 1375.

CHIMOR: DIVERSE WATER MANAGEMENT (850 TO C. 1470 CE)

The Chimor state arose in the Moche valley around 850 CE. Just like those of the Sicán, its first rulers may have been descendants of Moche nobles; they were also heavily influenced by their contemporaries, notably the Wari. Over the next four centuries, the Chimu people expanded their economic and political authority over a broad region of the north and north-central Peruvian coast. They inherited a great deal from their predecessors, but with one major difference. From the beginning, the Chimu lords adopted a different approach when building their capital, Chan Chan.[8]

Chan Chan lies near the mouth of the Moche valley and developed into an enormous city, rivaling highland Mexico's Teotihuacán from earlier centuries. From the beginning, this was a horizontal metropolis whose rulers focused on providing adequate food supplies. No one knows exactly how large the urban population became. By 1200 CE, the city sprawled over twenty square kilometers. Some 26,000 artisans lived in mud and cane houses along the southern and western edges of the central precincts. They included metalworkers and weavers. Another 3,000 lived close to the royal compounds, with about 6,000 nobles and officials occupying detached adobe compounds nearby. The rulers themselves remain anonymous to us, as they left no written records, but they lived in nine secluded and walled compounds in the heart of the city. Each had its own water supply, lavishly decorated residential quarters, and a burial platform for when the enclosure became the ruler's sepulcher.

Oral traditions and seventeenth-century Spanish chroniclers tell us that a ruler named Michancamán led Chimor at the time of the Inca conquests between 1462 and 1470 CE. Apparently, his courtiers had well-defined ranks, among them the "preparer of the way," an official who scattered powdered shell dust where the ruler was about to walk. Each leader built his compound near the others but inherited no wealth whatsoever. This institution, commonly known as "split inheritance," forced Chimu leaders to acquire additional territory, wealth, and taxable subjects by conquest. They also adopted the practice of forcibly moving conquered peoples to regions far from their homelands, as did the Inca.[9]

Chimor became a hierarchical, highly organized society with carefully defined classes of nobles and commoners, as well as a strict legal system used to enforce the social hierarchy. Trusted officials governed different areas of Chimor's domains. In political terms, the state was highly successful. At its height, the Chimu ruled over a kingdom that ranged beyond the northern coastal zone of the old Moche kingdom, extending into the south and along some 1,000 kilometers of coastline.

The lords maintained their growing state through a combination of force and tribute. They soon realized the importance of efficient communications, based on a road system that connected each valley. Many of the roads were little more than tracks, but they linked every area of Chimor. This was vital, for the state's tribute and material wealth flowed to the center. As in other ancient civilizations, the lords were careful to reward loyalty and bravery in battle with insignia and valuable gifts. They were also well aware that their entire state depended on food supplies that could not be acquired solely by force or through tribute.

For centuries, coastal farmers like those of the Moche had used highly flexible agricultural systems, located along coastal hillsides where they could maximize spring water and runoff from rainstorms. This strategy worked fine when population densities were relatively low. In stark contrast, confronted with rapidly growing urban complexes and fast-rising populations, the Chimu invested heavily in highly diversified and closely organized water management and agriculture.

Chan Chan itself relied heavily on step-down wells, many drawing on the high water table close to the Pacific. Low-lying terrain east of the city allowed for an elaborate system of sunken gardens with a high water table that extended five kilometers upstream from the Pacific. By 1100 CE, tax laborers had dug a huge canal network that watered flatlands to the north and west of Chan Chan. Irrigation water also replenished the city's aquifer. When a strong El Niño in the same year diverted the course of the Moche River and seriously damaged irrigation systems upstream of the capital, the rulers boldly embarked on a seventy-kilometer-long canal to bring water to the devastated fields from the neighboring Chicama valley.[10] This ambitious project was never completed, partly because the city expanded into areas upstream where the water table was much deeper. In the end, the city contracted toward the Pacific and the shallower coastal water table.

This was only part of Chimor's remarkable warfare against drought and ENSOs. The lords' plans were ever more ambitious and expensive.[11] They created elaborate canals throughout the kingdom to bring water to different parts of potential fertile river valleys. Some were up to forty kilometers long. The Jequetepeque valley north of the Chicama has fertile agricultural land both on the floodplain and on adjacent, irrigable desert plains, as well as plentiful marine resources at the coast. The valley's north side still bears signs of at least four hundred kilometers of canals built over many centuries. This vast canal system was never in use all at one time, for there was not enough water to fill its entire length. The communities that depended on the canals must have developed carefully organized delivery timetables for equitable delivery for all. When one realizes that today's local farmers water their crops about every ten days, one has an impression of the logistical complexity of the system. For all its intricacy, the Chimu canal works provided a practical way of mitigating extreme climatic events, such as heavy ENSO rainfall, and some means of responding to political uncertainty caused by water shortages.

FARMING LANDSCAPES AND TWELVE VALLEYS

The south side of the Jequetepeque valley is a different story, with large coastal sand dunes extending as much as twenty-five kilometers inland. A severe drought between 1245 and 1310 CE created dunes so extensive that a large settlement at Cañoncillo was abandoned in the late fourteenth century, when encroaching dunes covered fields, blocked irrigation canals, and buried houses. Such longer-term desertification was on a far larger scale than the havoc caused by aridity and heavy rainfall. One could repair such damage, but encroaching sand dunes were beyond human intervention. One could only move elsewhere.

Aridity was one thing; too much rainfall from El Niño events was another. Local leaders and water experts at larger Chimu urban centers like Farfán Sur, Cañoncillo, and others built elaborate overflow weirs as part of their irrigation canals, especially for the aqueducts that bridged deep ravines. The weirs could slow water flow and prevent erosion. The same aqueducts had stone-lined conduits that allowed water to flow through the base of the structure without damaging it. These helped somewhat, but there are signs that many of them were rebuilt after being overwhelmed. Another strategy: building crescent-shaped stone sand brakes in areas near the coast. These slowed the flow of dune sand into irrigation canals and fields, but many proved ineffectual.

The Moche had relied on individual communities, with few attempts to centralize management of agriculture. The damaged villages simply moved and built a new canal system. This was all very well when population densities were low, despite intense competition for the most fertile land. The Chimu lived in a much more densely populated farming landscape. They developed large towns and cities and practiced agriculture on a regional scale. They invested in entire agricultural landscapes that they created with huge investments of tax labor. These organized farming landscapes included large storage reservoirs and the terracing of steep hills to control cascading downslope water. The biggest investment was in long canals that carried water from deep-cut riverbeds to terraces and irrigated land far away, even during severe droughts. This was investment for the long term and enabled Chimor to create thousands of hectares of new fields, which they planted and harvested two or three times a year. Previously, only one harvest had been possible, coinciding with the annual flood from the mountains.

Eventually, land reclamation became uneconomic, even with huge labor forces. Instead, the lords of Chimor turned to conquest, rationalized by the institution of split inheritance, where each ruler had to acquire his own farming land. In the end, they controlled more than twelve river valleys with at least 50,500 hectares of cultivable land, all worked with simple hoes or digging sticks. Agriculture on this scale required efficient, ruthless administration. It could be no other way, given the enormous investment in construction and management required. The rulers restricted individual movement, forced many subjects into cities, and exercised highly centralized control of food supplies and population. Such centralized administration had strategic advantages, for the Chimu could respond to lengthy droughts or major ENSO events on a regional rather than local scale. They could divert crops from one area to another, bring undamaged irrigation canals on line, and deploy large numbers of workers to repair flood damage.

Chimor relied on long-term planning and achieved agricultural miracles in an environment where only 10 percent of the land was cultivable. Fortunately, the kingdom could also rely on the anchovy fishery. According to historical sources, the fishers were people apart who traded food with the farmers. The shore dwellers were virtually immune from drought but not from El Niños, when upwelling off the coast slowed and anchovy catches plummeted.

While Maya lords led their subjects into an environmental chaos, Chimor's elite survived prolonged droughts and exceptionally powerful ENSO events during the Medieval Climate Anomaly. Their leaders supervised an elaborate organized oasis based on massive human labor and draconian, centralized control. They also relied on a rigid social order and rituals that mediated between the natural and supernatural worlds. Their environment preadapted both leaders and farmers alike to the harsh realities of one of the driest environments on earth, where rainfall was rare and water came from afar. Droughts occurred in everyone's lifetime; the state adapted by diversifying food supplies, conserving every drop of water, and relying on fishing to widen the food base. The hard-won experience of the ancestors, expert opportunism, and long-term planning paid rich dividends.

Chimor controlled its subsistence, but the lords could not dominate the watersheds that nourished their kingdom with mountain runoff. The scale of their farming operations was now so large and complex that they had trouble administering water supplies, especially upstream. In about 1470 CE, Inca conquerors from the highlands gained strategic control of the watersheds and overthrew the state. The kingdom became part of Tawantinsuyu, the Inca "Land of the Four Quarters." Farming and irrigation continued, while the new lords of coastal valleys moved Chimor's expert artisans to Cuzco in the highlands.

The coastal states thrived on various scales because they had an intimate knowledge of their environment and of the water sources that nurtured their soils. Their leaders and farmers lived in river valleys where severe droughts were routine and El Niño clusters ravaged their farmlands. They were well aware of the telltale signs of an impending ENSO: reduced anchovy catches, a southward-flowing inshore current, unfamiliar tropical fish, warmer water close inshore. Everyone, whether Moche, Sicán, or Chimu, could foretell potential catastrophe, as well as drought in the highlands caused by ENSOs that reduced runoff at planting time. Of all the Andean states with their diverse social responses to climatic and environmental changes, it was Chimor that realized long-term planning helped maintain sustainability, which endured into Inca times and later.

Maintaining sustainability has always been a challenge along the Peruvian coast and in the Andes. One is struck at once by the diverse ways in which small communities adapted to local conditions and unpredictable droughts, which sometimes lasted a generation or more. The river valley farming along the coast practiced by the

Moche and Chimu could never have thrived without careful long-term planning by cultivators in small communities. The emphasis on "future-proofing" for times of harsh drought is especially notable with Chimor, where the state invested heavily in hydrological works like canals that linked valleys.

Both the Moche and Chimu were hierarchical societies, which gave their leaders access to tribute in the form of forced labor as taxation. Clearly this worked on the basis of a social contract between leaders and commoners, where everyone perceived advantages in careful water management and looking ahead at potential risk. In retrospect, this seems to have been more closely organized in Chimu society, but however effective the leadership was, agricultural expertise—local knowledge of the environment—and community-based labor were quite clearly of central importance. So were the kin ties that acted as glue between farming villages close to irrigation works. Communal cooperative labor was all-important. Centralized, authoritarian governance deployed labor, but local knowledge and kin ties held things together. Add to this equation the anchovy fisheries close offshore, and there was sufficient diversification in food supplies to feed people in arid years.

Compare this to Tiwanaku in the highlands, where food surpluses depended on rainfall and also irrigation schemes that were ultimately community based. When long-term drought took hold, the centralized authority of the rulers at Tiwanaku dissolved inexorably, and the state fell apart. In the countryside, local communities, with their close kin links, survived.

THE ASTONISHING HIGHLANDS: TIWANAKU (SEVENTH TO TWELFTH CENTURIES CE)

The altiplano (Spanish for "highland") is close to the southern margins of the Quelccaya ice cap, which means its ice cores are a sensitive barometer of climate change. Lake Titicaca lies just 120 kilometers south of Quelccaya, and sediment cores from the lake provide a second source of accurate information on rainfall. So the question is, how did people in the past react to the changes in climate documented by ice cores? Fortunately for archeologists, Tiwanaku, one of the largest known pre-Colombian sites in South America, lies close to the shore of Lake Titicaca.

Tiwanaku became a major city-state between the seventh and early twelfth centuries CE.[12] According to the ice cores, these five or so centuries were a period of generally warm and relatively wet conditions. There were drier intervals, but the climate was relatively stable. The ice cores include layers of wind-blown sediment from an extensive network of gridded, raised field systems that surrounded the city. We know of some 19,000 hectares of such fields in Tiwanaku's hinterland alone. In the city's heyday, the state's agriculture depended on these field systems created and maintained by village communities. The most productive lay at strategic points on the altiplano, in plots surrounded by irrigation canals. Even the muck from the surrounding canals provided rich nourishment for the fertile raised soils—as did dung from the llama, the main local domesticated beast. When rainfall was plentiful, high water tables and canals would soak the fields, providing not just ample water but also excellent protection against frost damage to growing crops. This drenching was particularly relevant to the success of the most prestigious crop, maize. The Tiwanaku farmers also harvested potatoes—a staple of commoners but, again, easily decimated by frost on high ground—and plots of ulluco, with its brightly colored potato-like root and edible spinach-like leaves. Raised-field agriculture was so productive that the villagers developed widespread grids of such gardens between the seventh and early twelfth centuries. Local field systems eventually became carefully integrated regional systems, providing the food surpluses that supported a political elite, a complex ideology, religious beliefs, and widespread trade in the lowlands and into desert landscapes.

Perhaps 20,000 people lived in these agricultural "suburbs" that surrounded the politico-religious heart of Tiwanaku. Its city was monumental, rich with imposing edifices. A great sunken court, the Kalasasaya, dominates an earthen platform faced with stones. Nearby, a boundary row of upright stones delineates a rectangular enclosure, with a nearby gateway carved with an image of an anthropomorphic god, sometimes called Viracocha. Smaller buildings, enclosures, and huge statues lie near the ceremonial structures, part of a powerful iconography that features condors and pumas, along with anthro-

The sunken court at Tiwanaku, Bolivia. *Alvar Mikko/Alamy Stock Photo.*

pomorphic gods attended by lesser deities or messengers. The center of Tiwanaku was an intensely sacred place, ruled by semidivine lords whose names are unknown. This elite ruled from the summit of a carefully organized kingdom supported by herding and subsistence agriculture so extensive that archeologists can still detect the furrowed remains of the long-abandoned raised fields that surround the city.

Compelling economic and political forces lay behind the facade of the highland state, much of its prosperity based on locally smelted copper and other trade around the southern shores of the lake and with the distant coast. Informal trade networks using llamas linked the highland city with a colony at Moquegua on the Pacific watershed about 325 kilometers away. Such colonization was no coincidence, for the two centers lie in the heart of a fertile maize-growing environment. Site surveys led by Charles Stanish and others in the southwestern Titicaca basin located this city and two other large towns with strong cultural links to Tiwanaku in the same southern valley.[13] Numerous people lived there over the centuries. Some of them traveled widely. Over 10,000 individuals with strong highland Tiwanaku connections lie in a large cemetery at the contemporary city of Chan Chan near the coast.

The trade between central Tiwanaku and the peripheral sites appears to have been relatively informal but involved the flow from the periphery of commodities and goods unobtainable at the core of the state. Unlike the later Inca, people at Tiwanaku made no effort to maintain a formal road system. But they did sustain colonies at lower altitudes, peopled by inhabitants who kept close, long-term relationships with parent communities on the highlands. Most of the trade lay in the hands of kin groups and merchants with strong people-to-people ties along the well-established trade paths. The llama pack trains may have numbered in the hundreds—they are much smaller today—and, judging from modern observations, traveled between fifteen and twenty kilometers a day. This trade spread the state's ideology, expressed in clay vessels and art, which strengthened Tiwanaku's economic and political authority over an enormous area of the highlands and lowlands. The commerce even continued after the Tiwanaku state disintegrated.

FLUCTUATING HOT AND COLD

Earlier we noted how Tiwanaku emerged during centuries of relatively warm and stable climate with more rainfall than previously. Just as happened with the Maya, agriculture expanded, raised fields proliferated, and population densities rose. These were the good times, when Tiwanaku experienced a burst of construction and expansion as the prestige and spiritual influence of its rulers dominated a wide expanse of both the altiplano and the distant, arid coast. It was not to last.

The Quelccaya ice cores and Lake Titicaca borings reveal that a severe drought ravaged Tiwanaku and its domains in about 1000 CE.[14] The amount of rainfall declined precipitously; Lake Titicaca's water level dropped by more than twelve meters after 1100. The lake shore receded significantly for kilometers, stranding many hectares of raised fields. Meanwhile, local water tables fell far below their normal levels in previous centuries. Many hydraulic systems that had so artfully sustained the surrounding canals became useless, especially inland from the lake.

The radical landscape changes occurred at a time of rising, increasingly dense populations. Areas of formerly marshy terrain ideal for intensive agriculture became drier environments. Despite both engaging in far less intensive farming and growing more diverse crops at short notice, the people were unable to create the bountiful food surpluses of earlier times. Within a few generations, the large-scale, carefully organized farming systems created and managed by the state's rulers were no longer viable. The agricultural system that had supported Tiwanaku's very foundations fell apart. Severe drought caused the Tiwanaku state to implode in generations of economic, political, and social instability. An increasingly fragmented, competitive agricultural and herding economy developed, which had inevitable and serious political and social consequences.[15] Successful local leaders in better-watered areas acquired independence from a state whose rulers had long invoked their powerful divine ancestry and associations. These changes took hold between 1000 and 1150 CE.

Just as was the case with the Maya, the dissolution of the state was a complex and patchy event. People continued to reside in segments of Tiwanaku and in a nearby major agricultural area in the Katari valley into the twelfth century. Ritual practices continued uninterrupted. The traditional lifeway survived during what seems to have been a long and chaotic process of disintegration.

Quelccaya's ice cores tell us that droughts continued to plague the region, with an especially long one during the thirteenth and fourteenth centuries, as well as during a period of irregular warming around 1150 CE (which coincided with the Medieval Climate Anomaly, the European warm snap, as we will see in Chapter 11). During this unseasonal heat, Tiwanaku and another great Andean state to the north, the Wari, finally dissolved economically and politically. By this time, communities had dispersed from valley floors and lower valley slopes to higher altitudes where water was thought to be easier to obtain.

With raised fields no longer viable, many of these communities occupied previously unexploited and uninhabited local environments ignored by Tiwanaku. The effects on highland society were dramatic. In the once prosperous Katari valley, farmers moved into numerous smaller villages that were a quarter the size of those in Tiwanaku's heyday. Gone was the carefully nurtured social hierarchy of earlier centuries and what must have been sometimes slavish devotion to the political and ritual activity, including feasting, that had tied people to the now abandoned city. Survival meant moving away from the once rich agricultural landscapes near Tiwanaku to readily defensible locations at higher elevations nearer to sources of glacial water. By 1300 CE, hilltop forts were commonplace; skeletons discovered by archeologists document violence, perhaps endemic warfare.[16] After five centuries of uninterrupted raised-field farming and crowded urban centers, savage aridity made centuries of agriculture around Lake Titicaca unsustainable. No densely populated towns thrived on the altiplano and adjacent highlands for centuries until the Inca Empire took control of the region in the mid-fifteenth century.

Raised-field cultivation effectively ceased until modern times. It was "rediscovered" by two archeologists, American Alan Kolata and Bolivian Oswaldo Rivera, who investigated abandoned raised fields about ten kilometers north of Tiwanaku.[17] Their excavations cut across raised fields and neighboring canals, as well as through occupation mounds, to reveal the measures taken to improve drainage and resurface the fields with mud from the canals. Joined by archeologist Clark Erickson, local farmers, a group of agronomists, and others, they launched a project to bring back the old way of farming. Together, they constructed a precise replica of a raised field, using only traditional tools such as foot plows. The new plot proved highly successful and also confirmed that small family and kin groups could easily build, cultivate, and maintain such fields. A subsequent program of controlled experimentation has led many altiplano farmers to adopt the long-lost raised-field farming system that once supported an entire civilization.

Once again, we learn that traditional agricultural knowledge is still highly relevant in today's world. Unfortunately, much of this knowledge is vanishing before we can put it to use in our warming world. We fail to heed the lessons of the past at our peril.

8

CHACO AND CAHOKIA

(c. 800 to 1350 CE)

Pine Island Sound, Florida, c. 1100 CE. The canoe slips quietly through a narrow defile in the mangrove swamp and emerges into open water. A long fiber line and a pole thrust into the bottom set the craft in place. Husband and wife cast a fine net and let it settle, waiting patiently. The net feels heavy to their touch and moves slightly. They haul the struggling pinfish aboard and move on. But the paddle grazes the bottom. The paddler makes a mental note and moves into deeper water. Depths change constantly here in the recent cooler weather, so everyone is relying more heavily on whelks and other edible mollusks.

The Native American Calusa people of what is now southeastern Florida flourished in a low-lying maritime environment, subsisting on a wide variety of fish and mollusks. Everyone relied on watercraft, living in compact, permanent settlements, for higher ground was rare and mobility a challenge. Food supplies were plentiful and reliable, but the sea level was never stable. A rise or fall of a few centimeters could decimate a sea grass fishery or destroy oyster or whelk beds. Storing food was almost impossible, so each isolated village relied on canoes to maintain contacts with others in a society where trade and reciprocity helped everyone. Ultimately, the glue that held everyone together was the intangible: their empirical knowledge and their supernatural beliefs, reflected in a complex ritual life.

The realm of the intangible lay at the core of ancient North American life. From the beginning, over a span of more than 15,000 years,

Homo sapiens adapted successfully to a very broad array of environments in North America: from harsh arctic tundra, to temperate forests, to the desolate, arid landscapes of much of the West. Native Americans passed down the secret of these adaptations and the vast reservoirs of knowledge associated with them by word of mouth over hundreds of generations. Much of this knowledge that helped people cope with climate changes of all kinds still survived intact into the nineteenth century. Much still remains, embedded in chant and song and in unwritten knowledge that is carefully cherished and rarely shared with others. The major vicissitudes of global climatic shifts—the constant interactions between the atmosphere and the oceans, El Niños, major drought cycles, and the warming that triggered major sea level rises, and so on—were the background to innumerable successful, and unsuccessful, *local* adaptations firmly based on traditional experience and knowledge. It is only now that we are beginning to realize that sustainability and resilience to these shifts were major players in Native American history in modern-day Canada and the United States.

We can only describe a few examples in these pages, but they represent a cross-section of the potential for dramatic advances in our knowledge, with important relevance to today's debates over future climate change.

DROUGHTS AND FISHERS (1050 BCE TO THE THIRTEENTH CENTURY CE)

The endless seesaws of sea surface temperatures in the tropical Pacific bring both drought and rainfall to California in unpredictable abundance. For thousands of years, hunters and gatherers living on the coast and in the interior adapted to drought or flood with familiar strategies.[1] They swayed with the climatic punches, fell back on permanent or reliable water supplies in arid years, and consumed less desirable foods if necessary. Many groups relied on different forms of oak trees for easily stored, nourishing acorns. Fishing societies along the Southern California coast exploited natural upwelling in the Santa Barbara Channel and relied on anchovies as a staple, along with acorns.[2] As in other hunter-gatherer societies, people "managed" their environment by burning off dry grass to enhance new growth or to attract game.

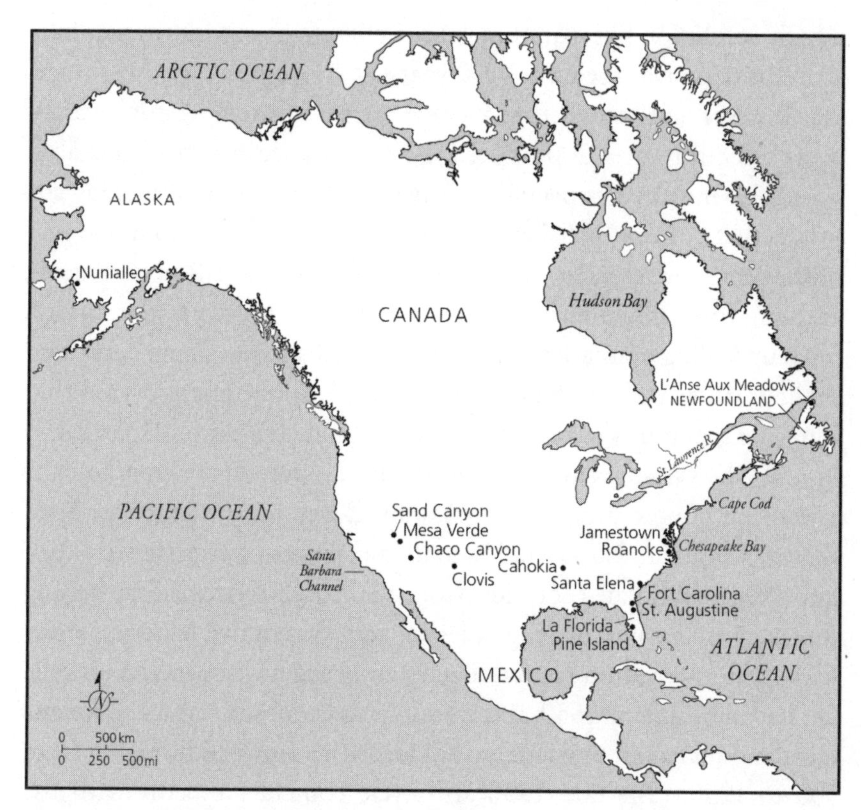

North American sites in Chapters 8 and 13.

Many groups retreated into swamps and wetland environments when droughts arrived. As always, conservative strategies that minimized risk, combined with flexibility and opportunism, ensured survival in diverse arid and semiarid landscapes.

People like the coastal Chumash fishers of the Santa Barbara region could cope effortlessly with short-term climatic variability such as El Niños. Longer-term shifts were another matter, detected today from deep-sea and lake cores and tree rings. Fortunately the Santa Barbara Channel has yielded a 198-meter deep-sea core, 17 meters of which cover climatic changes since the Ice Age.[3] The foraminifera (plankton and other similar simple organisms) sedimentation accumulation rate was rapid, which was ideal for studying a highly sensitive environmental setting. Using the foraminifera and radiocarbon dating, a father-and-son team, Douglas and James Kennett, acquired a high-resolution portrait of maritime climate change at twenty-five-year intervals over the past 3,000 years.

The Kennetts found that average sea surface temperatures varied up to 3°C. But after 2000 BCE, the climate became more unstable. From the human perspective, life became more complicated, as the productivity of coastal fisheries could vary dramatically from one year to the next. The intensity of coastal upwelling was a critical barometer, marking times when nutrient-rich colder water moved to the surface. Such upwelling dramatically improved local fisheries. Using deep- and shallow-water foraminifera in the core, the Kennetts discovered that water temperatures were relatively warm and stable from 1050 BCE to 450 CE. Warmer surface water reduced natural upwelling, so fisheries were poorer. From 450 to 1300 CE, sea temperatures dropped sharply, to about 1.5°C cooler than the median since the Ice Age. For three and a half centuries, from 950 to 1300, marine upwelling was particularly intense, engendering very productive fisheries. After 1300, water temperatures stabilized and became warmer. The upwelling had subsided by 1550. Interestingly, the cool sea surface temperatures and increased upwelling coincided with regional droughts from 500 to 1250. (The 800–1250 arid cycle coincides broadly with the Medieval Climate Anomaly.) Similar drought cycles appear in Sierra Nevada mountain tree ring sequences at many locations in the American West, with two prolonged droughts in one sequence lasting for more than 200 years and 140 years, respectively, megadroughts by any standards.

The Chumash peoples of the Santa Barbara Channel and their ancestors have long been famous for flourishing in what was erroneously called a "Garden of Eden," with rich fisheries close offshore and bountiful acorn harvests on land. But even in cycles of good years with abundant rainfall and good catches, many communities lived from year to year. The cooling after 450 CE improved the fisheries, but there were many more people to feed. Then the eight and a half centuries that followed saw frequent drought cycles, which may not have caused much trouble on the coast but hit hard inland. As populations rose, territorial boundaries became closely delineated. Chief vied with chief for control of territory and oak groves and fought over permanent water supplies. We know of occasional malnutrition, also of traumatic wounds, from cemetery skeletons dating from about 1300 and 1350, when bows and arrows first appeared. Outbreaks of local conflict, short-lived and violent, were inevitable in climatically stressed groups in areas of unpredictable rainfall, highly localized food supplies, and intense political and social competition.

Chumash society changed profoundly after 1100 CE, when violence and persistent hunger—even, perhaps, local population crashes—were endemic. Settlements grew larger; people lived closer together; hierarchies of large and smaller settlements developed as hereditary elite lineages led by chiefly families developed powerful mechanisms for controlling trade, resolving disputes, and distributing food across landscapes where food resources varied dramatically in locations only a few kilometers apart. Dances and other rituals validated the new social order through an association known as *antap*, which served as a social mechanism that linked powerful individuals over considerable distances. Thus, the Chumash survived until Spanish contact in the sixteenth century in a volatile political milieu where cooperation ensured survival in a challenging natural environment.[4] The Chumash example shows that carefully controlled ritual can foster sustainability and resilience in societies where food supplies were not necessarily abundant.

Chumash fisheries benefitted greatly from natural upwelling during the Medieval Climate Anomaly between the ninth and thirteenth centuries CE. So did two major societies: Chaco Canyon in the Southwest and Cahokia in the Mississippi River's American Bottom near modern-day Saint Louis. Although some 1,500 kilometers apart—theories differ as to whether they were aware of each other—both rose to prominence, then disintegrated between the twelfth and thirteenth centuries. Their lifespans coincided with the Medieval Climate Anomaly, with its somewhat warmer and wetter climatic conditions over less than fifteen short-lived generations.

CHACO CANYON: A CLIMATIC TAP DANCE (C. 800 TO 1130 CE)

The San Juan basin covers much of northwestern New Mexico and adjacent parts of Colorado, Utah, and Arizona.[5] This is a landscape of broad plains and valleys. Small mesas, buttes, and short canyons define the basin. Chaco Canyon is a spectacular ceremonial center and earthwork complex, famous for its nine multistoried "great houses," or large pueblos. More than 2,400 other archeological sites, large and small, lie within and around them. For more than three hundred years, between 800 and 1130 CE, this area was home to a surprisingly dense farming population whose members lived cheek by jowl, with daily life characterized by a constant murmur of voices from the terraces and plazas and an undercurrent of scents and odors, including those of sagebrush, human sweat, and rotting food. They maintained a sustainable agricultural system in a marginal region for agriculture with very unpredictable rainfall of about two hundred millimeters a year, with wide variations. Ultimately, everything depended on meticulous water management.[6]

The Chaco core, today known as the Chaco Canyon National Monument, lies halfway down the canyon. Here, the most famous pueblos flank a seventeen-kilometer length of the Chaco Wash that flows sporadically through the canyon. Of all the "great houses," Pueblo Bonito—a semicircle of rooms pressing on a central plaza with once underground circular ceremonial rooms, or kivas—is the best known.[7] Each great house was a vibrant place, where factionalism and social tensions were routine. The site itself came into being at what may have been a sacred spot, marked by striking rock formations in the nearby canyon walls. Pueblo Bonito also lay opposite the conspicuous South Gap, which funneled summer storms into the canyon's heart.

Initially, between 860 and 935 CE, Pueblo Bonito was a small masonry settlement, a modest, arc-shaped place, but one with strong spiritual associations. The inhabitants dwelt in a layered world: the sky, the earth, and the underworld. Their village was the *sipapu*, the point of emergence from the underworld. Elaborate rituals revolved around the summer and winter solstices and the passages of the sun and moon. The Pueblo world was always one of harmony and order, its basic values reenacted in dramatic performances. The group mattered more than the individual; people focused on maintaining a human existence that had always remained the same and would be the same in the future. Chaco life revolved around maize agriculture and religious

Pueblo Bonito, Chaco Canyon, New Mexico. *Roberthardingl/Alamy Stock Photo.*

belief in a landscape where the harsh realities of an arid environment defined human existence.

But other factors were also at play in an increasingly complex, more politicized era when more emerging leaders became hungry for increased power and religious authority. By 1020 CE, Pueblo Bonito had potent spiritual associations. Building resumed after 1040. Within thirty years, Bonito had acquired a labyrinthine, indeed mesmerizing, complexity. It had started as a residential settlement but then became a great house, a ceremonial building with powerful ritual and political associations and plenty of storage space but few permanent inhabitants, though visitors crowded it at the solstices and other major events.

Chaco's farmers relied on a variety of water-management systems to nourish their crops. They cultivated the wash's floodplain and slope washes from the cliffs and relied on floodwater cultivation when there was enough rainfall. A battery of scientific methods, including airborne LiDAR survey, excavation into long-filled canals, sediment coring, and study of water sources using strontium isotopes, tell us that

Chaco farmers used a wide variety of water sources obtained by channeling runoff.[8] Complex systems of artificial channels and earthen canals formed part of a multifaceted irrigation system tailored to local conditions. Rapidly changing rainfall patterns and the unpredictable environment required nimble social responses to sudden abundances as well as to water scarcities through deployment of labor from the great houses and small settlements. Powerful ritual associations tied to the elite occupants of the great houses underscored both farming and water management. Descent through the female line, confirmed by DNA studies of burials at Pueblo Bonito, was a vital factor in the success of Chaco agriculture, for women had powerful voices in water management, as their ritual activities were tied to both fertility and to water.[9] Kinship, inheritance, and conservation of precious water supplies were all-important in a culture where women were powerful members of society and often ritual leaders. Pueblo Bonito's leadership was hereditary, religious, and powerful. The cultural order revolved around impermanent realities, such as unpredictable water supplies, and the sky, with its light and dark objects that showed up against the surrounding terrain.

Such centralization of Chaco leadership may have sustained a social system that constantly faced unpredictable environmental conditions and climate change. But it was lightly incised into the landscape. Social controls that ensured the monitoring of soils and changing water supplies, along with the deployment of labor at short notice, were essential elements in the risk management that lay behind long-term survival.

In the final analysis, Chaco society succeeded due not so much to its powerful leaders but to the flexible autonomy of the household, an autonomy guided by the belief that most labor was ultimately for the common good. No one could be self-sufficient in an arid environment like the San Juan basin, which is one major reason why elaborate ceremonial observances held society together. The observation of solstices and other major ceremonial events commemorating important moments in the agricultural year brought people together in an environment where kin ties and obligations extended far beyond the boundaries of the canyon. It could not be any other way in a society where reciprocal ties, sometimes reflected in pottery styles, linked kin groups living many kilometers apart. Such ties came into play when food was plentiful in one place, limited in other. In times of scarcity, you moved to stay with fellow kin in better-watered places in the expectation that they would do the same if fortunes changed. Cooperation, mobility, and resilience went hand in hand in a society affected by climate shifts that changed their tempo from year to year. The relationship between the Chacoans and the climate was like an intricate dance, a minuet of partners—the farmers and the endless gyrations of rainfall, temperatures, and growing seasons. The climate set a quick, agile pace. Its human partner had to be nimble, quick to respond to hints from the land and sky, or else the dance would end in disaster. And the Chacoans were adept responders.

Four major weather patterns affect the San Juan basin and the Colorado plateau. The moist polar Pacific airmass arrives from the northwest, brought by cyclonic storms that move to the south and southeast. This reverses during summer, when warm, moist tropical air from the Gulf of Mexico brings rainfall, interspersed with occasional incursions of warm Pacific air that produce more rain. Mountain uplift brings sometimes massive, localized summer thunderstorms mostly between July and early September. But only a small amount of rain falls annually, with considerable variation from year to year. Everything depends on the movements of air masses thousands of kilometers away and on local topography. The amounts can vary dramatically at locations only a few kilometers apart.

Summers are hot, winters cold, with a growing season over the basin of about 150 days, but as much as a month shorter in lower locations like Chaco Canyon. The people who dwelt in the canyon were at the mercy of capricious and often unexpected climatic shifts. Short-term global events such as El Niños also had a profound effect on agriculture from one year to the next.

The Chacoans were probably barely aware of long-term climate change, for the living generation and its ancestors enjoyed the same basic adaptation, which we could call a form of "stability." But every Chaco farmer was only too aware of shorter-term, higher-frequency changes—year-to-year rainfall shifts, decadelong drought cycles, seasonal changes, and so on. Droughts, El Niño rains, and other such fluctuations required temporary and highly flexible adjustments, such as farming more land, maintaining two to three years' worth of grain reserves, relying more heavily on wild plant foods, and also moving across the landscape.

The strategies worked well for centuries, so long as the Chacoans lived sustainably, farming the land at well below the number of people per square kilometer it was capable of supporting. However, when the population increased to near carrying capacity, the people became increasingly vulnerable to El Niños and especially to short or more prolonged droughts. Even a year of poor rainfall and bad crops or torrential rains could destroy a household's ability to support itself within weeks or months. Longer dry cycles were potentially catastrophic.

Tree ring dating is a fundamental climatic proxy in the Southwest. Today we have a year-by-year tree ring record for Chaco Canyon from 661 to 1990 CE and data from other proxies that show us that waves of great house construction coincided with periods of more plentiful rainfall, further indicating the stable climate allowed for increased populations. Building surged at Pueblo Bonito and elsewhere between 1025 and 1050, with three periods of greater than average rainfall separated by short dry spells. The San Juan basin was a precarious farming environment even in the best of times, but a higher water table than usual and more rain made the canyon one of the more secure places. Between 1080 and 1100, two decades of aridity gave the farmers problems, fortunately offset by the high water table. Then abundant rains returned, when construction again accelerated, not only at Chaco but also in the northern San Juan basin at places like Aztec and Salmon Pueblos.

By 1130, Chaco's inhabitants were so dependent on domesticated plants that they were ill-prepared for the fifty-year drought that began in that year with only one short break. Maize yields fell catastrophically; wild plant growth was stunted. Nor were rabbits or other wild animals readily available. After 1100, the people had imported turkeys from the northern San Juan basin as a substitute, but these did not satisfy the need for more food supplies. Had the drought lasted just a few years, both the great houses and smaller communities would have survived. But after 1130, there seemed to be no reprieve from severe drought, and people were hungry. The Chacoans resorted to an ancient strategy: they moved elsewhere.

Mobility had always been commonplace at Chaco. Families had long moved in and out of the canyon. They would arrive or leave for a season, decide to dwell with kin living at a distance in the highlands, or solve a long-festering dispute by moving elsewhere. The long-established communities of which they were part continued to flourish, each with its own gardens and water supplies and with rights over other foods and resources. When the fifty-year drought arrived, there was no great mass exit; nor did hundreds, if not thousands, of Chacoans perish in a traumatic famine. Instead, people moved away, family by family or sometimes in larger kin groups. They walked to areas with more plentiful rainfall, to communities where they had nourished ties of kinship and trade for centuries.

Emigration from Chaco during the twelfth century began as the usual irregular, constant flow of families in and out of the canyon. But as conditions worsened, efforts to grow more crops failed. The water table sank. Eventually small-scale movements became rivulets of households moving to growing communities elsewhere. The increasing depopulation of Chaco Canyon reached a point where entire long-established communities upped and left. Only a few persistent hamlets hung on, until they were no longer viable. We can only glean a few hints of the suffering of those who stayed. For instance, archeologist Nancy Akins's bone research disclosed that by the eleventh century, 83 percent of Chacoan children suffered from severe iron deficiency anemia, which increases the risk of dysentery and respiratory disorders.

So long as rains continued, fresh gardens could be planted and new settlements constructed for their owners. After 1080 CE, when the rains faltered, there was still a frenzy of construction. But at some point the leaders of the great houses lost control of the elaborate ritual apparatus that brought precious exotica and commodities like timber beams to places like Pueblo Bonito. No longer could they orchestrate the meticulously performed ceremonies that had marked the cadence of the farming year for many generations. Chaco was no longer the spiritual center of the world. People gradually dispersed elsewhere. The beating heart of the Ancestral Pueblo moved northward to the San Juan River, southwestern Colorado, and the Mesa Verde region. Chaco Canyon became only a memory, but a powerful one, incised into the oral traditions of dozens of Pueblo communities to the north.

Chaco's history had always revolved around relationships with others, with people, fellow kin, and communities outside the narrow confines of the canyon. There was what we can call a Chaco world, based on great houses, that had become an increasingly important religious center. Chaco's leaders never controlled outlying communities, but different parts of this world tied themselves to the canyon in distinctive ways with widely differing objectives, both for themselves and Chaco's leaders.

It's nonsense to talk of a collapse at Chaco caused solely by drought, just as it's misleading to describe a Maya breakdown in the same terms. The Chacoans had always lived in a marginal farming environment with all the vulnerabilities surrounding sustainability that this entailed. For generations, Chaco's leaders were enabled by a period of above-average rainfall and highly successful farming. This required other communities to legitimize their role. And when Chaco faltered and abandonment ensued in a complex series of events, the center of the canyon's world shifted north. That anything survived is because of powerful ancestral memory, a belief that the gods controlled not only the cosmos but also humans. But, like people, the deities had obligations to share their bounty with others—the ancient notion of reciprocity. Chaco was based on three unspoken values: harmony, flexibility, and movement. These same principles lay at the heart of many ancient societies, which were keenly aware of the importance of resilience, sustainability, and risk management. We still have much to learn from these sound approaches to daily existence.

MOVED BY DISASTER (1130 TO 1180 CE)

The drought of 1130 to 1180 CE brought Chaco's great houses to their knees. As Chaco declined, political influence shifted northward to the leaders of the Aztec and Salmon Pueblos.[10] Two clusters of farming population now achieved some prominence, one of them centered around Totah, north of Aztec, the other around Mesa Verde in the Four Corners region. A wave of great house construction ensued there lasting about sixty years. Any form of large-scale building ceased around 1160, when beams from the central Mesa Verde area reveal a slowdown in tree felling during the next great drought.

Unlike Chacoan farmers, northern San Juan communities relied entirely on dry maize farming mainly at elevations over 1,829 meters. In arid years, loose-grained soils could support many more people than actually lived in the region, even during severe droughts. During the tenth century, the local people lived in small, dispersed communities, often five to ten residential units with a kiva and storage rooms. But settlements grew larger and farming populations less dispersed from the late 1100s to the 1200s. Many previously small hamlets grew into villages with multiple room blocks—though not to the extent of Chaco's great houses.

Population growth continued into the mid-1200s, with numerous independent communities competing for farmland, control of trade routes, and political power. Raiding and warfare became endemic as many groups moved into defensible locations in canyons. This was the era of the Cliff Palace and other famous Mesa Verde pueblos in deep canyons, but those in the nearby Mancos and Montezuma valleys prospered off large drainages. Here, people tended to congregate close to the most productive land, where they could survive extremely severe droughts if there were no restrictions on mobility or access to the best soils. Population densities rose from 13 to 30 people per square kilometer to as many as 133 three centuries later. Village sizes doubled. But once population densities approached the carrying capacity of the land, and all the most productive soil was under cultivation, adapting to long drought cycles became much harder.

Sand Canyon Pueblo near modern-day Cortez in southern Colorado became one of the largest fortified communities in the northern San Juan, placed near a canyon head with its ample water sources. Between 1240 and 1280 CE, as many as seven hundred people lived behind a huge enclosure wall. Between eighty and ninety households occupied Sand Canyon's room blocks, in a pueblo that they built, occupied, then abandoned over a brief span of forty years. Unlike Pueblo Bonito, this was more a residential site than a ritual center, although feasting and solstice ceremonies were part of the annual round.

In 1280, after forty successful years, Sand Canyon's inhabitants experienced a more severe drought than any in their collective experience. This was the moment of climatic truth. Refined dendrochronology, along with reconstructions based on the Palmer Drought Severity Index, have provided detailed environmental information. Meteorologist Wayne Palmer developed an algorithm that uses data on rainfall and temperature to measure dryness. His index is widely used to measure long-term drought today and in earlier times. A lattice of reconstructed climate change, information on soils, potential crop-production data, and available wild foods shows that the thirteenth-century drought did not destroy the carrying capacity of the Sand Canyon landscape. Thus, a reduced population could have remained in the region throughout the worst of the aridity.

The tree ring research required is complex and demanding. For example, most current sequences use cool-season moisture conditions, which will give way to curves based on studies of spring and summer rainfall. The fir trees from Mesa Verde provide some of the strongest climatic signals, their rings yielding data on the climates of the previous autumn, winter, and spring. Using sophisticated correlation analyses, the researchers have reconstructed the amount of September–June rainfall down to individual decades over the past 1,529 years. We now know that there were several prolonged cool-season droughts in the Mesa Verde region during the twelfth and thirteenth centuries. The 1130–1180 arid cycle brought severe aridity during both the winter and the warmer months. To add climatic insult to injury, drought conditions during the early summer prevailed over a large tract of the region during the entire century. The early and late thirteenth century were the most severe of all. As had happened in Chaco Canyon a century earlier, population movements out of the region began, slowly occurring over many decades until the area was empty at the close of the thirteenth century.

Ultimately, a process of dispersal unfolded in the northern San Juan, just as it had earlier in Chaco Canyon. In both places, the Ancestral Pueblo followed centuries of tradition and moved away from drought-stricken land, a process that saw war, suffering, and a gradual movement of farmers southeast into the Little Colorado River drainage, the Mogollon highlands, and the Rio Grande valley, areas where there was little rainfall change. The Native American tribes we know today as Hopi and Zuni are the descendants of the Ancestral Pueblo people who moved to this area.

Movement was one solution to the overpopulation of marginal farming landscapes. But today's Southwest has witnessed a population explosion and the exponential growth of major cities like Phoenix, Tucson, Las Vegas, and Albuquerque. Such huge cities and large-scale agriculture have placed major stresses on groundwater and other scarce water supplies as global warming intensifies, long-term droughts become more commonplace, and moving people to areas with more reliable water supplies is no longer an option. Again, long-term planning and thinking about water supplies will be vital for a much drier and more densely populated future. The classic strategy of migration that may have served earlier civilizations well is no longer a viable option in our time.

MISSISSIPPIANS (1050 TO 1350 CE)

The environmental conditions in the Mississippi River valley were very different from those in the Southwest. The Mississippi is a stupendous river by any standard, with a huge triangular drainage area that covers about 40 percent of the United States, exceeded only by the Amazon and the Congo. A capricious river, the Mississippi can deliver catastrophic floods and prolonged periods of low water causing drought. The fertile, humid American Bottom, as the river floodplain near Saint Louis is called, was a major focus of human settlement long before Europeans arrived.

Cahokia presided over the American Bottom from about 1050 CE, a magnificent indigenous ceremonial center, the political and ritual focus of what archeologists call a powerful "Mississippian kingdom."[11] The great center was both a ritual landscape and a thriving urban and ceremonial complex that extended across both banks of the Mississippi. The population of the densely occupied and fortified central area of Cahokia, with its imposing earthen mounds, rose rapidly from somewhere around 2,000 people to 10,000 to 15,300 inhabitants in the half century between 1050 and 1100 CE. Many of them were immigrants from elsewhere in the central United States during the Medieval Climate Anomaly, a period of slightly warmer climatic conditions over wide areas of the world between about 800 and 1250 CE.

The Native American chiefs who directed this magnificent center depended on kin ties, deft political maneuvering, long-distance trade monopolies, and the supernatural powers attributed to individuals believed to have close connections to the spiritual world. Slippery alliances held together Cahokia's loosely knit domains, tied together by transitory bonds of loyalty, personal and kin relations, and an ancient cosmology that embraced a cosmos with three levels, the uppermost and lowermost of them inhabited by powerful supernatural beings. One of them was the mythic Birdman, an avatar of warriors. The Great Serpent, a great denizen of the underworld, was in constant opposition to the Birdman. The political influence and spiritual tentacles of this cosmology extended from the Gulf of Mexico coast to the Great Lakes and into numerous Mississippi tributaries.

Cahokia was a unique, local Native American adaptation to favorable environmental conditions, rising and increasingly diverse population densities, and the need for larger food surpluses to support an ever-more elaborate chiefdom. The American Bottom had fertile soils ideal for maize agriculture and was rich in both fish and waterfowl, but the risk factor was high with such a dense population, especially in a sequence of bad years. Agriculture was firmly under elite control and extended onto nearby uplands, as many farmers moved to higher ground to avoid an increasingly high water table and periodic river flooding. Without the upland farmers, the 10,000 or more people living in the central precincts would have faced a heightened risk of food shortages. Nor did they have the flexibility to adapt to more severe climatic conditions.

Calibrated tree ring readings from a drought measurement network that covers North America tell part of the climatic story. As in the Southwest, the half century between 1050 and 1100 CE was relatively wet. These were the years that the upland population grew rapidly. Then came drought, starting with a fifteen-year dry cycle after 1150. Frequent arid periods dominated most years, the 1150 dry cycle coinciding with the major drought in the Southwest discussed above.

Getting ancient population figures is usually a problem, for site counts can be misleading. However, confirmation of the above shifts comes from Horseshoe Lake, an oxbow of the Mississippi just north of Cahokia.[12] It has yielded two important cores that provide a 1,200-year record of fecal stanols—organic molecules that originate in human guts and, amazingly for us, survive in sediments for centuries and millennia. They are a proxy for measuring population changes over time. A high ratio of coprostanol (stanol produced by human dung) to 5a-coprostanol microbes produced in the soil indicates that there were large numbers of people on the landscape. Low ratios reflect much smaller populations. Changing stanol ratios from the two lake cores record rapid population growth in the tenth century, with a maximum in the eleventh. By the twelfth century, the Cahokia watershed population was in decline, with a population minimum in about 1400.

Maize, which grows during the spring and early summer, was a decisive staple during Cahokia's heyday. When the population decline took hold around 1150, temperatures and fecal stanol ratios declined in unison until the thirteenth century. But a major flood event in 1200, the first in five centuries, also inundated cultivated land, grain stores, and numerous floodplain settlements.[13] Such major inundations typically occur in spring and early summer, during the critical maize growing season. Serious crop failures and food shortages must have ensued. The great flood must have reshaped Cahokia, for the leaders could not draw on the large numbers of people needed to clear debris and drying sediment from farmed land or to rebuild shrines and houses. While many Cahokians may have moved to live with kin on higher ground, the damage was done. This was at a time when we know that slow depopulation was already under way, defensive palisades were under construction, and the construction of elaborate public buildings had slowed.

Cahokia's leadership, perhaps comprised of several elite families, imploded as the American Bottom population dispersed. By 1350, Cahokia was deserted except for a few small villages. The inhabitants scattered; entire precincts melted into the soil; earthworks and mounds vanished under woodland.

Numerous elements formed part of the religious beliefs that revolved around Cahokia. They included the realities of life and death, pervasive substances like water, and such phenomena as changing moonlight and darkness, as well as their own 18.6-year moon cycle. This was manifest in spectacular monuments, earthen mounds, mortuary shrines, and elaborate public mortuary rituals involving, among other things, an avenue to the dead built of earth and running through standing water. Sweat houses centered on shrines also played a role in Cahokia's ritual life. These were all powerful experiences, but they depended on people's belief that Cahokia was the center of the world and on their loyalty.

But as at Chaco, the imprint of the kingdom and its infrastructure was relatively light on the landscape. When this leadership faltered, perhaps because of a blinkered, narrow perspective that focused too strongly on the spiritual realm, order was disrupted—until new, much smaller, local centers rose to prominence. Disputed succession, a less than charismatic leader, a successful revolt—all could have toppled those who ruled Cahokia and were probably among the events that did. The social bonds that linked the people to the elite broke down. The diverse population of immigrants and locals became disillusioned and abandoned the American Bottom. Judging from the striking lack of oral traditions about Cahokia, those who left must have been profoundly alienated. Cahokia vanished from history for seven centuries until noted by early-nineteenth-century archeologists. The American Bottom was not entirely depopulated, however: semisedentary maize farmers and bison hunters, who were much more mobile than their predecessors, settled there.[14]

Mississippian chiefdoms like Cahokia depended on maize agriculture as well as political and social leadership that relied on tribute paid to powerful chiefs. These leaders maintained the loyalty of their followers and lesser centers by redistributing tribute and through commonly held religious beliefs and elaborate rituals, such as sweat lodge ceremonies. This was a classic way of organizing hierarchical societies, but one that suffered from fatal weaknesses. Everything depended on kin ties and loyalty, the latter always a fickle quality in societies where factionalism was often endemic, as was the case in southwestern pueblos. The experience of Chaco Canyon and Pueblo Bonito is a textbook example of the fundamental power of kin and community in a society where kin obligations extended far beyond the narrow confines of the canyon. Here ritual obligations revolved around agriculture and the passage of the seasons, between water and drought. Even the relatively authoritarian role of the leading women and families in the great pueblo depended ultimately on ancient ties of kin and doctrines of mobility in the face of climate change long after Chaco Canyon was all but deserted. Pueblo villages are embedded in their environments and the social relationships that entangle and maintain them through the generations right into modern times.

Centralized authority depended on long-term stability and reliable food surpluses. In many kin-based societies, taut control, even through force, could only extend over a relatively limited territory, perhaps as small as fifty kilometers across. This was certainly the case with Cahokia, whose influence and power were based on trade and complex religious beliefs. When major floods and droughts affected the American Bottom, the Cahokian chiefdom disintegrated as lower temperatures took their toll. Economic and political changes rippled across the Mississippi valley. Disputes erupted into warfare; the great center was palisaded, as were other settlements to protect against unrest. Major population centers imploded as their inhabitants dispersed in response to fighting both internally and with neighbors in competition for authority and power. In the Mississippi valley, control of maize surpluses and more exotic commodities cemented political power. When Spanish conquistadors traveled through the Southeast, they encountered not a unified, powerful chiefdom but dozens of fortified villages and towns, often separated by deserted territory. Some local populations numbered in the thousands, with hints of significant wealth that spoke to Spanish greed. But they found themselves entangled in a morass of enmities and rivalries. Survival and sustainability for these Native American societies depended on complex political and social realities, which were a far cry from those of highly centralized states like Angkor, which we visit in the next chapter.

9

THE DISAPPEARED MEGACITY

(802 to 1430 CE)

Wealth, beauty, and magnificence: Angkor Wat in Cambodia is a stupendous architectural masterpiece, said to be the world's largest pre-twentieth-century religious building. A power-hungry ruler, Suryavarman II, built his palace-temple at the zenith of the Khmer empire during his reign between 1113 and 1149–1150 CE. Its size is overwhelming. Just the main temple, with its lotus-bud towers, covers 215 by 186 meters and rises more than 60 meters above the surrounding land. Its moated enclosure encompasses an area 1,500 by 1,200 meters. Angkor Wat makes the temple of the Egyptian sun god Amun in Karnak or the Cathedral of Notre-Dame in Paris look like a village shrine.[1]

Angkor Wat is close to the Mekong River, which floods between August and October. The overflow fills a nearby lake, the Tonle Sap, which swells to 160 kilometers long and a depth of up to sixteen meters. As the waters recede, thousands of catfish and other species lurk in the shallows, making it one of the richest fisheries on earth. What is known as Greater Angkor lies between the Tonle Sap and the well-watered Kulen Hills. The surrounding plains allowed Angkor Wat to expand in every direction with plenty of space for rice cultivation. Reservoirs and canals distributed water over thousands of hectares of farmland and supported the spectacularly wealthy Khmer civilization, which prospered between 802 and 1430 CE. There was, however, one intractable problem: it was almost impossible to maintain crop yields in a region where they were never abundant without careful water management. Even with plentiful crop yields, growing population densities increased the risk of food shortages. Angkor's leaders had but one option: clear yet more forest and plant more fields to feed an inexorably rising population.

Compact urban centers had a long history of over six centuries in Southeast Asia, especially in the Mekong Delta. During the eighth and ninth centuries CE, these gave way to more dispersed cities, which reached their peak during the thirteenth century. A series of ambitious Khmer kings fashioned a powerful and more stable empire. The rulers created a cult of divine kingship and elaborate centers dominated by extravagant shrines, as at Angkor Wat and nearby Angkor Thom. Thousands of commoners labored for a state in which absolutely everything flowed to the center. In 1113, King Suryavarman II started building Angkor Wat, using a carefully organized workforce from throughout his domains, fed with fish from Tonle Sap and enormous rice harvests.[2]

Every detail of Angkor Wat reflects some element of Khmer mythology. Khmer cosmology consisted of Jambudvipa, a central continent with Mount Meru, the cosmic mountain, rising from its center. The sixty-meter central tower imitates Meru; four other towers represent the lower peaks. The enclosure wall replicates the mountain range that was said to encircle Jambudvipa, with the surrounding moat representing the Sea of Milk where gods and demons churned ambrosia.

Angkor Wat and Angkor Thom were redolent with cosmic and religious symbolism, including astronomical observatories, royal burial places, and temples. Generations of researchers have studied their art and architecture, but the dense vegetation that mantled both sites and the entire area hampered any attempts at systematic field survey. In 2007, an international team got together to launch a state-of-the art project using an array of cutting-edge techniques to understand what was really going on at Angkor Wat and its wider landscape. This revolutionary work has revealed that the temple complex at Angkor Wat was far larger and more convoluted than previously expected. But perhaps even more sensationally, the team employed airborne LiDAR technology, a remote sensing method that uses light, in the form of

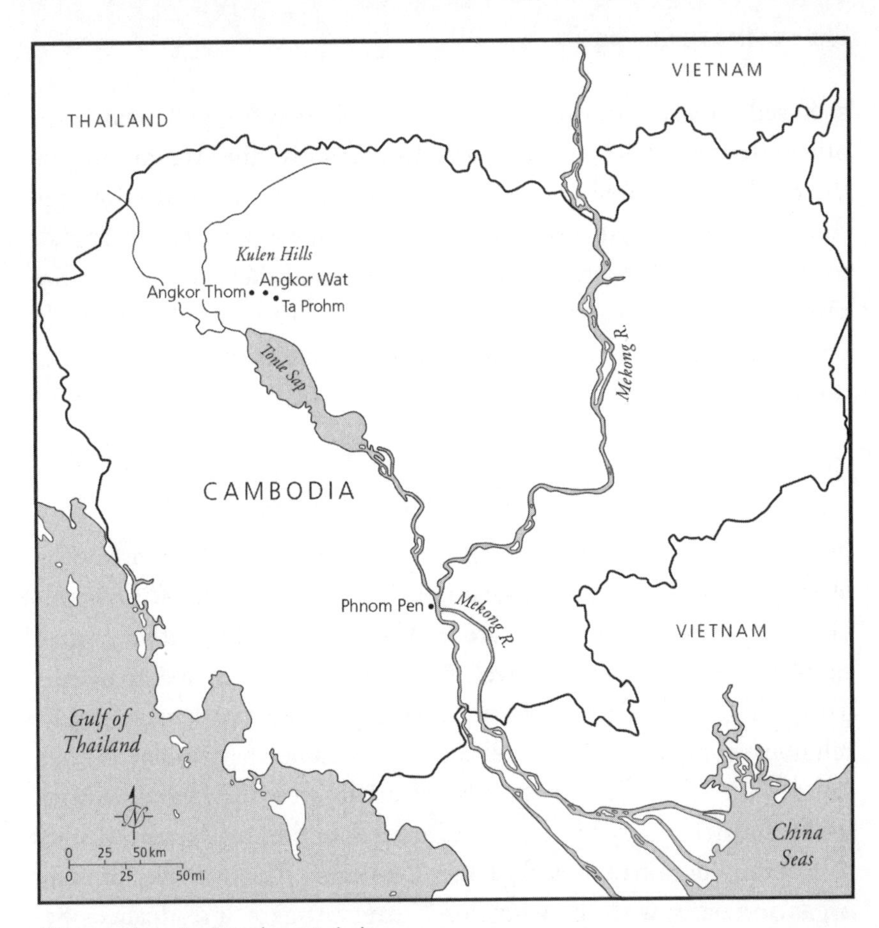

Khmer sites in Southeast Asia.

Angkor Wat, Cambodia, from the air. *Sergei Reboredo/Alamy Stock Photo.*

a pulsed laser, to measure the variable distance from the drone (or other airborne vehicle) to the earth. In the case of the Angkor project, the resultant captured images have allowed the team to "see through" the dense jungle that surrounds the main temple complex at Angkor to reveal the unexpected. In an academic bombshell, the project has revealed the presence of a lost "megacity": a vast network of roads, ponds, canals, thousands of rice fields surrounded by narrow banks, house mounds, and over 1,000 small shrines.[3]

UNBRIDLED MAGNIFICENCE

Greater Angkor's rural-urban sprawl stretches over a huge area of at least 1,000 square kilometers, supporting perhaps 750,000 to 1 million people (the figure is debated). Meanwhile, only relatively small numbers (perhaps 25,000) dwelt within the temple enclosure of Angkor Wat itself. The relationship between the wider settlement and its religious-political-economic elite center is somewhat similar to that, say, of New York City to the central urban area of Manhattan with its St. Patrick's Cathedral or of Greater London to its central City of London, dominated by St. Paul's Cathedral. This was the ultimate organized oasis, with the whole of Greater Angkor spread across extensive, organized rice fields. Archeologically, it is reminiscent of the densely settled landscapes that once surrounded Maya religious centers, like Tikal and Caracol, also recently surveyed with LiDAR, but the scale here was much larger.

The great beating heart of Angkor Wat was not unique. You'll recall that while Angkor Wat was established by Suryavarman II (r. 1113–1145/50), it was in fact finished by King Jayavarman VII (r. 1181–1218). And this same softly smiling king (or so he appears in his sculptures) built another temple complex, Angkor Thom, aptly meaning the "Great City." This would be the final and most enduring capital city of the Khmer Empire. Lying some 1.7 kilometers north of Angkor Wat, Jayavarman's new city covered 9 square kilometers and housed between 30,000 and 60,000 people in its central area. About a half million more lived in suburbs that extended as much as fifteen kilometers from the center. This was no spur-of-the moment project. Rather, the LiDAR surveys indicate that the Khmer must have had

Angkor Thom long in mind, for they constructed a road grid half a century before the temple was even built. The grid tied the entire hinterland of the temples into a canal and road network that extended into the sprawling neighborhoods where most people lived.

Everything depended on adept water management. Long before either Angkor Wat or Angkor Thom, the Khmer had begun building *barays*: huge rectangular reservoirs to store water and for surplus water to be drained into the vast Tonle Sap Lake, which led to the Tonle Sap River and then on to the Mekong. By the ninth century CE, *baray* construction was in full swing, the foundations of an unchanging water-management system that formed a huge artificial delta. There were intake channels at the north end and a fan of channels to the south that flanked the East and West Barays close to Angkor Wat.[4]

This ingenious and flexible system allowed officials to move water in almost any direction across the level plain to conserve it or to drain into the great Tonle Sap Lake. Don't confuse this with the centralized irrigation systems developed in Egypt or Mesopotamia. While in all these lands the basic irrigation technology was simple and dependent on abundant human labor, only the Angkor state could marshal the scale of labor required to construct the major canals or reservoirs that were the lifeblood of Angkorean civilization. Australian archeologist Ronald Fletcher has aptly described this as "a risk management system that worked to mitigate the uncertainties of monsoon variation in a landscape of largely rain-fed, bunded rice fields."[5] At a basic level, he is certainly correct, and he also rightly called it a paradox. The Khmer created a versatile system that could handle unpredictable monsoon fluctuations. But they confronted a serious long-term problem. The large scale of the waterworks and their management made them difficult, almost impossible, to modify and even maintain when rapid changes were needed in the face of major climatic shifts.

Angkor's canal and bund network watered fields in the north and provided adequate supplies to ensure that the bund fields of southern Angkor produced high rice yields. The West Baray close to central Angkor serviced a relatively small area of the entire plain. It helped provide supplies for about 200,000 people, an adequate supply for dry years caused by monsoon failures. The system worked well until the late twelfth century, when water engineering focused more closely on the urban center. New channels—built, at least in part, to support the increasing numbers of people needed to manage and maintain major temples—ensured that water flowed through the center of Angkor. King Jayavarman VII alone doubled the number of temples in central Angkor during the late twelfth and early thirteenth centuries.

The resources required are mind-boggling. Just one temple staff member would have needed the services of about five farmers to produce his or her rice. Jayavarman VII's Ta Prohm temple (completed in 1186) and Preah Khan temple (completed in 1191) alone had more than 150,000 support people, who had to live nearby. Just these two medium-sized temples consumed the labor of a fifth of Greater Angkor's population. Yet they seem to have managed it. Water buffalo abounded, fish were common, and vegetables were plentiful. The state maintained a glittering facade and kept order with draconian severity and religious zeal. Indeed, the king would not appear in public without a massive display of force. At the time, it seemed as though this unbridled magnificence would never end—until monsoon failures came into play.

CAPRICIOUS MONSOONS (1347 TO 2013 CE)

Angkor's harvests had always depended on the Asian monsoon.[6] The westerlies brought by the monsoon intensify as they move north into Southeast Asia and the South China Sea. The monsoon rains are at their height in August and September, bringing intense tropical cyclones to the Bay of Bengal. The summer precipitation that falls on Angkor comes from steady monsoon rains and from torrential downpours brought to land by major tropical disturbances, especially the tropical cyclones. By the time they reach Southeast Asia, their strong winds abate, but they dump large quantities of rainfall as slow-moving storm systems for as long as four days. Such disturbances and less powerful equatorial easterly waves contribute about half the summer rainfall over Southeast Asia.

Angkor's twelfth-century empire-building rulers may not have known it, but their kingdom was far more vulnerable than it had been even a century earlier.[7] The state had maintained high levels of rice production by the simple expedient of expanding the amount of land deforested for agriculture on a massive scale. Most of the Angkor landscape was now bunded rice fields, with only scattered tree cover. When heavy monsoon rain fell, intense runoff and the resulting uncontrolled erosion would have denuded soils. Add to that the deforestation of the uplands, and serious ecological consequences were in train. Aerial photographs reveal thousands of abandoned ancient rice fields across a landscape that was cultivated far more intensively than today.

Moreover, Angkor's infrastructure, built originally as a form of risk management, was over five hundred years old when the climate became less stable. Its last *baray* had been built a century earlier. Angkor's enormous foundation was aging, increasingly hard to manage effectively, and more and more convoluted. Fletcher's archeological team has excavated a failed dam that was rebuilt in the tenth or eleventh century. Everything had been fine when the city's population was much lower and the system was new. When there had been damage, it was promptly repaired, but no more.

What exactly happened? We can needle out answers thanks to a tropical cypress, *Fokienia hodginsii*, found in Vietnam. It provides a tree ring record of El Niño Southern Oscillation (ENSO) events and monsoons from 1347 to 2013 CE. The ring thicknesses coincide with cold La Niñas and warm El Niños.[8] During the fourteenth century, wild fluctuations between the two—represented by huge monsoons and severe droughts—came into play. In addition, excellent speleothem records from Dandak Cave in India's monsoon region and from Wanxiang Cave in northwestern China agree well with the tree ring records from southern Vietnam, especially those for the thirteenth and fourteenth centuries.[9] Together, the evidence shows that the thirteenth and fourteenth centuries were a period of major climatic instability in Southeast Asia, as well as China and Mongolia, with unpredictable fluctuations between exceptionally strong monsoons and major droughts.

At first, the Khmer system could handle periodic droughts, as it had done for centuries, but it was vulnerable. Its dams were apparently unable to deal with major floods. Excavations into its two principal reservoirs, the East and West Barays, have revealed that exit channels were blocked, some of them as early as the twelfth century. There were times when the East Baray did not fill, resulting in serious water shortages during the drier conditions of the early thirteenth century. Then came heavy monsoon downpours before the fluctuations stabilized in the sixteenth century. By this time, there was a labor shortage: not enough people were left to shunt water around.

Imagine a 150-year drought, followed by abrupt and extreme monsoons, impacting a massive but centuries-old infrastructure. So much water arrived that the geriatric network fractured; the elites had neither the ability nor even the desire to fix what must have been catastrophic damage to delicately crafted canal gradients. The effects were pernicious. Wrecked fields could no longer support Angkor's dense urban population. Sustainability was in ruins. Nor could the farmers who had supported temples and their staffs for generations continue to do so. The elite had lived the high life, with large, elaborate households that were no longer viable. No longer did the state's rulers and high officials have the ability or power to recruit labor for major engineering works to fix the system. The food surpluses at their disposal were inadequate.

For centuries the complex and very stable water system had supported other features, such as roads and fishponds associated with the rice fields. Both the protein supply and the staple rice crop came under pressure. When the water system failed upstream, the damage cascaded downstream, among other things disrupting the entire road network. Both waterborne and land transport moved not only food but all manner of other commodities and luxury goods to integrated markets across Angkor's domains. For instance, no fewer than 6 percent of domestic goods found in Greater Angkor are of Chinese origin. Rumors, fear, and intercommunity rivalries caused social chaos to match the environmental disarray.

DISINTEGRATION (THIRTEENTH CENTURY CE ONWARD)

The severe drought during the 1360s must have wreaked havoc with food supplies. By the late fourteenth century, parts of Angkor became unusable, and the temple economy was in tatters. Quite apart from floodwater damage, the raging waters would have scattered debris of all kinds over the landscape, clogging vital canals and disrupting the carefully organized landscape even more.

Not everything got wiped away. A checkerboard of intact embankments and fields survived among breached roadways and embankments. The elite had the option of staying with wealthy relatives elsewhere, relocating with their rulers to other centers, or moving to their upcountry estates. But the artisans who depended on them and the farmers who delivered crops to the elites were left on their own in a devastated hinterland. These everyday people were abandoned to hunger and malnutrition, as water flowed across the landscape outside the human-created waterways while canals and dams disintegrated. Without question, farmers and others struggled to restore Angkor Thom's water systems, but Angkor had no royal presence for almost two hundred years until the mid-sixteenth century.

On the face of it, the implosion of Angkor came about as a direct result of megamonsoons and extreme droughts that descended on the Khmer like hammer blows. While this might seem a case of direct cause and effect, the historical reality was more complex.

As ever, religion played a starring (or, in this case, damning) role. The Mahayana tradition of Buddhism had become the state religion under Jayavarman VII (1181–1218), the builder of Angkor Thom. These were the very years that the monsoon became progressively weaker and food shortages occurred. Both the elite and farmers had not only to deal with the crisis but also to seek an explanation for what was happening. Religion was evoked. A reaction against royally supported Mahayana Buddhism led to the destruction of images of the Buddha displayed on major temple walls between the late twelfth and early thirteenth centuries. This iconoclasm was almost certainly the people's powerful response to drought, manifested in a belief that other faiths might provide better responses to continuing aridity.

For years, scholars assumed that Angkor fell in 1431, sacked by the competing Siamese kingdom of Ayutthaya, now in Thailand, a major international trading center that became one of the largest and most wealthy cities in the East during the sixteenth century. But we now know this was not so: Angkor had become uninhabitable by that time. The elite would have already left, taking their wealth with them. This was a period of profound political, economic, and social change for the Khmer state. It had to confront southward movements by both Thai and Vietnamese groups, which severed long-established terrestrial trade routes and coastal access for the Khmer. During the fifteenth and sixteenth centuries, trade became more globalized. The coastal cities assumed ever-greater importance, while the old and very stable rice production of the Khmer interior faded. Maritime trade with Arab, Indian, Chinese, and other seafaring entities assumed ever-greater importance. Beset by climatic difficulties, the empire faded into relative obscurity as depopulation accelerated.

Moreover, new religious beliefs conditioned against the old ways. The area had long had close commercial ties with India. These long-established trade routes carried not only products but also ideas and faiths, among them Theravada Buddhism. After the thirteenth century, this became the state religion of Cambodia. The new doctrine de-emphasized the long-established practices of supporting great temples and their numerous caretakers. The economic consequences rippled through Angkor's population as the influence of the great temples declined from the thirteenth century. Three centuries later, Angkor Wat was little more than a pilgrimage center, although Angkor Thom was still in use. A modest population remained in the area, as the center of state power moved southward to the Quatre Bras region around modern-day Phnom Pen.

The demise of Angkor involved far more than just climatic shocks. Implosion did not involve conquest; rather, it was a transformation. Khmer leadership and the center moved southeast in a shift from the massive organized oasis with its rice production into an area nourished by natural annual flooding. Here farmers were more impervious to drought. The Mekong River carries far more water at flood time than can be retained within its banks. When the Mekong overflows the Tonle Sap River after the monsoon rains, the floodwaters inundate the surrounding landscape. They fill the almost 113,000-square-kilometer freshwater lake of Tonle Sap, sometimes even flooding it.[10]

What happened in the Khmer world is exactly what happened with the Maya and, as we shall see, with Sri Lanka. Between the ninth and sixteenth centuries CE, dispersed urban civilizations in tropical environments from Central America to Southeast Asia broke down, their foundations rocked by the uncertainties of food supply and the erosion of traditional political power. Powerful dynasties rose and fell; warfare became commonplace; some elites moved to new centers. Civilizations disintegrated, in large part because maintaining sustainability was beyond the capabilities of rigidly administered and centralized states headed by divine kings devoted to unchanging religious ideologies. A period of transformation invariably followed. The farmers, who had once supported dynasties of remote rulers, maintained sustainable agricultural practices, which they reconfigured to reflect new environmental realities. Urban centers became more compact, often on the peripheries of now-vanished states. The great cities of earlier times, like Tikal, Tiwanaku, and Angkor Wat, gave way to new economic realities and political alliances, many of them based on international trade. In their place small compact settlements flourished around the peripheries of the great hinterlands.

Everywhere in the subtropical and tropical worlds, water management was a critical part of sustainability. The challenges such societies faced were numerous: well-defined wet and dry seasons, monsoons that could bring torrential downpours, ENSO events, hurricanes or typhoons, and both short- and long-term drought cycles. Above all, unpredictable rainfall was a permanent challenge that could vary dramatically from year to year. With new generations of climatic proxies available, we now know for sure that climate change helped destabilize medieval social and political systems over an enormous segment of the monsoon world in Asia. Farmers in South and Southeast Asia and in both northern and southern China were, and still are, at the mercy of climatic forces far from their homelands. The untamed magnificence of earlier times was indeed a myth.

INTO SRI LANKA (377 BCE TO 1170 CE AND LATER)

As we saw in Chapter 5, Roman trade with the Indian Ocean and as far away as the Bay of Bengal brought even Chinese silk to the Mediterranean—as did the Silk Roads across Eurasia. The great driver was the monsoon winds, which assumed great importance during the second century CE. Both Rome and Constantinople prospered greatly during the fourth century. The latter became the center of the growing

Eastern Roman Empire. Reliable monsoon winds, which changed direction seasonally, linked Alexandria and the Red Sea with India's west coast and Sri Lanka. An insatiable demand for ivory, spices, and textiles fueled the trade and brought wealth to increasingly complex societies in Sri Lanka.

At the time, the Anuradhapura Kingdom, founded in 377 BCE by King Pandukabhaya, dominated Sri Lanka. Its capital was in the so-called dry zone of the island, at the site of Anuradhapura, an important political center and a major intellectual and pilgrimage hub for Theravada Buddhism, the same branch that would later come to dominate Cambodia.[11]

The people had to work out how to irrigate a land with seasonal rainfall between December and February. To conserve water for use in the dry months, they constructed sizable reservoirs and dams, requiring large numbers of workers. Farmers also created irrigation systems, relying on gravity and cascading water in Anuradhapura's hinterland.[12] The central precincts expanded as the number of local monasteries and pilgrims increased. Their reservoirs grew larger and larger, with the Nuwarawewa Lake covering nine square kilometers in the first century CE. Yet demand for water increased even further, so the elite built more substantial dams and major feeder canals. The 87-kilometer-long Yoda Ela (or Jaya Ganga) canal linked major reservoirs on higher ground to more reliable water sources. By any standards, this was an engineering masterpiece with a gradient of only ten to twenty centimeters per kilometer.

Anuradhapura's water supplies depended on major hydrological projects commissioned by the elite, while local communities and monasteries built and ran their own modest cascades. Everything worked well during the centuries of relatively stable monsoon regimes and ample rainfall. The monasteries imposed ideological control over the irrigation hinterland, creating a theocratic landscape, with monks serving as both religious and secular administrators.

Then came severe climate instability between the ninth and eleventh centuries, which brought warmer temperatures and prolonged droughts. Unpredictable rainfall had serious consequences, as it had at Angkor. In contrast, the fourteenth to sixteenth centuries saw lower temperatures and increasingly dramatic incidences of heavy rainfall and drought. During the eleventh century, archeology tells us, the number of sites within a fifteen-kilometer radius of Anuradhapura declined to a mere eleven settlements.[13] No one still lived in the urban core. Most monasteries in the center and on the periphery went out of business. Routine maintenance of reservoirs and canals ceased; many silted up. Only a few small communities practicing slash-and-burn agriculture survived until recolonization of the dry zone during the nineteenth century.

As temperatures increased during the eleventh and twelfth centuries and Anuradhapura declined, Polonnaruwa, the second-most ancient kingdom in Sri Lanka, rose to prominence. Founded by Sinhalense king Vijayabahu I in 1170 CE, it lay further inland and on higher ground where temperatures were less extreme. Vijayabahu's grandson, Parakramabahu I (r. 1153–1186), commissioned canals and reservoirs even more extensive than those at Anuradhapura. His Parakrama Samudraya ("Sea of Parakrama") encircled his city and acted as both a reservoir and a defense against attack. The king's lake covered eighty-seven square kilometers and was in fact three reservoirs connected by narrow channels at low water. Thousands of laborers built his lake by hand, their reward being a spiritual one. This artificial sea—it was nothing less—supported an elaborate rice irrigation system that covered 7,300 hectares and fed the dense urban population.

Anuradhapura's and Polonnaruwa's temples and monasteries played a central role in agriculture and water management. Both served as focal points for the great annual religious festivals of the year attended by thousands of people from the city and hinterland. Major public celebrations defined the seasons both at Angkor and in Sri Lanka. Like those of Maya lords, they served to remind everyone of the complex, unspoken social contracts that linked everyone, whether priest, ruler, or commoner. The organized oases, created by reservoirs that surrounded the great stupas, enhanced the religious authority personified by the shrines. This serene landscape gave an impression of permanence and stability. But, as at Angkor, increasing temperatures and severely reduced monsoon rainfall cycles in the thirteenth and fourteenth centuries played havoc with reservoirs at a time of growing population densities and the resulting increasingly intense agriculture linked to the need for growing food surpluses. In response, rulers moved closer to much-reduced reservoirs and sometimes embraced new religious beliefs. The social transition was profound as people adopted the familiar strategy of dispersal in the face of prolonged drought. The population of the great cities imploded, leaving them as mere places of pilgrimage.

ON TO THE CATASTROPHIC NINETEENTH CENTURY: GREAT CHINESE AND INDIAN FAMINES (1876–1879 CE)

The Han emperors, who ruled China from 206 BCE to 220 CE, at the same time as the Romans in Europe, established modes of imperial irrigation and water control that persisted, albeit much modified, into the twentieth century. They faced major challenges, both with the Huang He River in the north and the Yangtze in the south. The Han and their successors relied on thousands of workers to build dikes and manage flood waters. There was constant tension between the central government and local interests, especially over the building of major waterworks. By the nineteenth century, China had failed to maintain sustainability at a time of unusually severe El Niño activity.[14] Thousands of years of occasionally inspired irrigation work, often sclerotic bureaucracy, and regimented labor failed to control the abrupt and often violent cycles of the natural world.

A long history of intermittent and sporadically effective famine relief addressed food shortages when floods and drought cycles attacked what was at best marginal sustainability. The culminating misfortune descended on northern China when the monsoon rains failed between 1875 and 1877 for two years in a row. The ensuing drought and famine were far more severe than dry conditions and crop failures in India during the same decade. An inefficient administration in distant Beijing did almost nothing until 1876, when tens of thousands of refugees appeared in city streets as far south as Shanghai. Starving farmers ate grain husks, grass seeds, and anything they could scavenge. American missionary Samuel Wells Williams saw "people like spectres hovering over the ashes of their burnt houses, and making pyres for themselves out of the ruins of their temples."[15] Most district administrators, overwhelmed by the scale of the disaster, did nothing or executed the thousands of famine-driven bandits by starving them to death in cages.

In the end, more than ninety million people suffered from hunger in an area larger than France. Missionaries and foreign consuls were the only sources of information for the outside world. They reported that the dead lay in large pits. Human flesh was openly for sale in the streets of villages and towns. Eventually, companies that had made millions off the opium trade established a China Famine Relief Fund. Devout Christian supporters of the China mission saw famine relief as "a wonderful opening," but few of those converted retained their faith after the monsoon returned in 1878. Missionaries estimated that only 20 to 40 percent of the hungry received any relief. By the time the famine ended, many villages had less than a quarter of their prefamine populations.

India was also greatly affected by severe climatic changes in the nineteenth century CE. The early years of Queen Victoria's reign until the 1860s were climatically relatively quiet in the monsoon world as good rains persisted. Just as happened in Angkor centuries earlier, plentiful rainfall brought good harvests and population growth. Indian farmers short of cultivable land took up hectares of less fertile terrain that supported adequate crops during wet years but were at best marginal for agriculture much of the time. All seemed well at a time when the British Raj began exporting grain from India. Then came the great El Niño of 1877 to 1878, followed by an unusual wave of similar events for more than three decades, especially in 1898 and 1917. The most severe El Niño, that of 1877, began with a savage drought in 1876 and endured for three years. A strong high-pressure system built over Indonesia and delayed the monsoon. Drought ensued, bringing widespread brush fires in its wake. In 1877, the huge pool of warm water in the Southwest Pacific moved eastward, generating one of the most powerful El Niños in history. Mass fatalities overwhelmed much of the tropical world, especially among farming populations that depended on rainfall alone for their crops rather than irrigation.

Famine ensued in India, where the drought was the worst since 1792. The rains failed; crops withered. The British authorities refused to introduce price controls, which set off a wave of frenzied speculation. As grain riots broke out and many laborers starved, even in well-watered areas, millions suffered and died, even as the British continued to sell Indian rice and wheat on the world market.

The great famine was a human-caused disaster. Refugees flocked to cities, where the police turned them away, 25,000 people from Madras alone. Many perished, while others wandered aimlessly in search of food. Meanwhile, the British Raj argued that famine relief might save lives but would only result in more people being born into penury; thus it did not actively attempt to feed the starving populace. The official policy was laissez-faire, with at least 1.5 million souls dying in the Madras area alone. Thousands were too weak to plant when the rains returned. At subsidized work sites where starvation rations fed the laborers, the dead and dying lay on all sides, "cholera patients rolling about in the midst of persons free of the disease." Vigorous protests in the press and by a vociferous minority in the government were of no avail. Writer Mike Davis has argued convincingly that this disaster laid the foundations of Indian nationalism.

The 1877 cataclysm was the first time that many colonial governments truly had to confront a commonplace but often dismissed elephant in the climatic room: near-universal famine and hunger in the lands they were exploiting, to which the locals' only clear solution at the time was out-migration. No one was ready for the massive population dispersals that resulted, which foreshadowed the mass migrations of the late twentieth century and today. Subsistence farmers, deeply attached to their ancestral lands, ran out of familiar options and adopted the only survival strategy open to them: dispersal in search of food and places where they could grow crops.

The great nineteenth-century famines in China and India rendered basically dismissive centralized governments virtually powerless. In the case of India, the British Raj was more concerned with profiteering from world grain prices than with the plight of local farmers. Their interventions caused chaos on an egregious scale. Millions perished in an era before the international relief organizations of modern times. The self-serving, centripetal Khmer state expanded rapidly and was overwhelmed by the maintenance demands of hydrological schemes that required massive human labor, meticulous organization, and efficient, decentralized administration. Like the

Maya and the farmers of Tiwanaku, the most effective solutions lay in transformed societies based not on imposing cities and shrines but on self-sustaining rural communities.

Herein lies a relevancy for our world. Millions of subsistence farmers and poverty-stricken people now suffer from food insecurity. Drought and hunger in the Democratic Republic of the Congo, South Sudan, Zimbabwe, and the Sahara's Sahel region are now virtually endemic, while a third of Afghanistan's population, some eleven million people, suffer from food insecurity. The story is nuanced and complex, but just as the colonial-era West carved up the world, fighting over land and resources and dehumanizing the people, so the battles and exploitation of people and resources often continue. Faced also with inefficient and often corrupt and uncaring local bureaucracies, the only survival strategy for people becomes out-migration, just as it was in the past.

Thousands of famished late-nineteenth-century Chinese and Indian villagers moved in desperate searches for nourishment and reliable water. Today, ecological migrants number in the tens of thousands or more in a warming world where drought is rapidly becoming endemic. And yet we in the West build walls (metaphorical or otherwise) against those very people whom we exploit. What gives us that right? Our economic system hardwires us to be this way, for capitalism has within it strong notions of corporate profit and exploitation. Consequently, different governments are compelled to protect their lands and their resources, and to push others down. Is this the best way, as a species, to deal with the issues of a warming world?

We now live, of course, in an era of cities with populations in the millions. But this causes us to forget the lessons of the past: we forget the massive investments in self-sustainability and cooperation in many rural communities, and we forget their long experience of risk management. If we cooperated with such communities, learned from them, and shared our capital by investing in their ways of living and dealing with issues, this would be a far more potent investment in the human future than ever-lavish military expenditures and the like.

10

AFRICA'S REACH

(First Century BCE to 1450 CE)

"The railway line was a mass of corpses." The descendants of the victims still call it *Yua ya Ngomanisye*, "the famine that went everywhere."[1] The drought in central Kenya lasted from 1897 to 1899 and decimated the small, autonomous Kamba and Kikuyu societies on the eastern side of the Great Rift Valley. In some places, crops failed three years in a row. In an earlier era, the farmers might have had sufficient reserves to survive, but these were now colonial times. The Uganda Railroad was under construction. Precious grain expropriated from nearby communities passed to laborers working on the tracks. Bubonic plague, probably imported from India by migrant workers, descended on the region, killing thousands. Hungry locals took to raiding. The railroad police burned their villages in retaliation, destroying even more food. Lions and other carnivores stalked humans in daylight; hyenas ate victims collapsed by the roadside. The British authorities made some desultory attempts to feed the survivors, but the losses were already enormous. To the west in Uganda, losses from the famine exceeded 140,000 people.

Famine descended after some years of excellent harvests and plentiful rainfall that had led to population growth concentrated in crowded settlements. Just as they did in medieval Europe, farmers took up marginal lands to grow more food. Local and long-distance trade prospered as the good rains persisted. Then a drought of truly biblical proportions followed in 1896, caused by a major El Niño, succeeded by a La Niña in 1898, and another El Niño in 1899. The Ethiopian Highlands, once the lush crucible of the Aksumite civilization, were now so dry that the Nile flood was the lowest since 1877–1878. Severe drought settled over East and Southern Africa, as well as the Saharan Sahel. Millions of farmers from Mount Kenya southward to distant Swaziland experienced serious crop failures. The misfortunes multiplied. Ceaseless rinderpest outbreaks attacked cattle herds, smallpox ravaged communities, numerous locust swarms descended, and other evils persisted in the face of a major climate shift. Simultaneously, European imperialists were encroaching. The British took advantage of the suffering by expanding their new protectorate based in Nairobi to annex much of the Kamba and Kikuyu territories. To the south, Cecil John Rhodes took possession of what was to become Rhodesia. The famed spirit mediums of the Shona divine Mwari at Great Zimbabwe proclaimed, "White men are your enemies . . . [R]ain clouds no longer visit us."[2]

Climatic shifts and other disasters transformed African society. Dynamic village economies dissolved as trade of all kinds disintegrated, crop diversity shrank, and crop yields tumbled. Power passed from traditional chiefs to puppet leaders appointed by colonial governments. African societies were now on the lowest rungs of a ladder of power firmly tied to Western-controlled global markets for grain and raw materials. Social inequality and underdevelopment, justified through scientifically absurd racist ideologies, became the norm as the European "scramble for Africa" took hold.

MASTERING *BASADRA* (BEFORE 118 BCE TO MODERN TIMES)

In 916 CE, Arab geographer Abu Zayd al-Sirafi wrote, "The *basadra* [summer monsoon wind] gives life to the people of the land, for the rain makes it fertile, because, if it didn't rain, they would die of hunger."[3] Centuries later, in 1854, American meteorologist Matthew Fontaine Maury published his *Explanations and Sailing Directions to Accompany the Wind and Current Charts*.[4] He used hundreds of ship observations to reveal the circulation of the Indian monsoon. In 1875, the Indian Meteorological Service came into being as an attempt

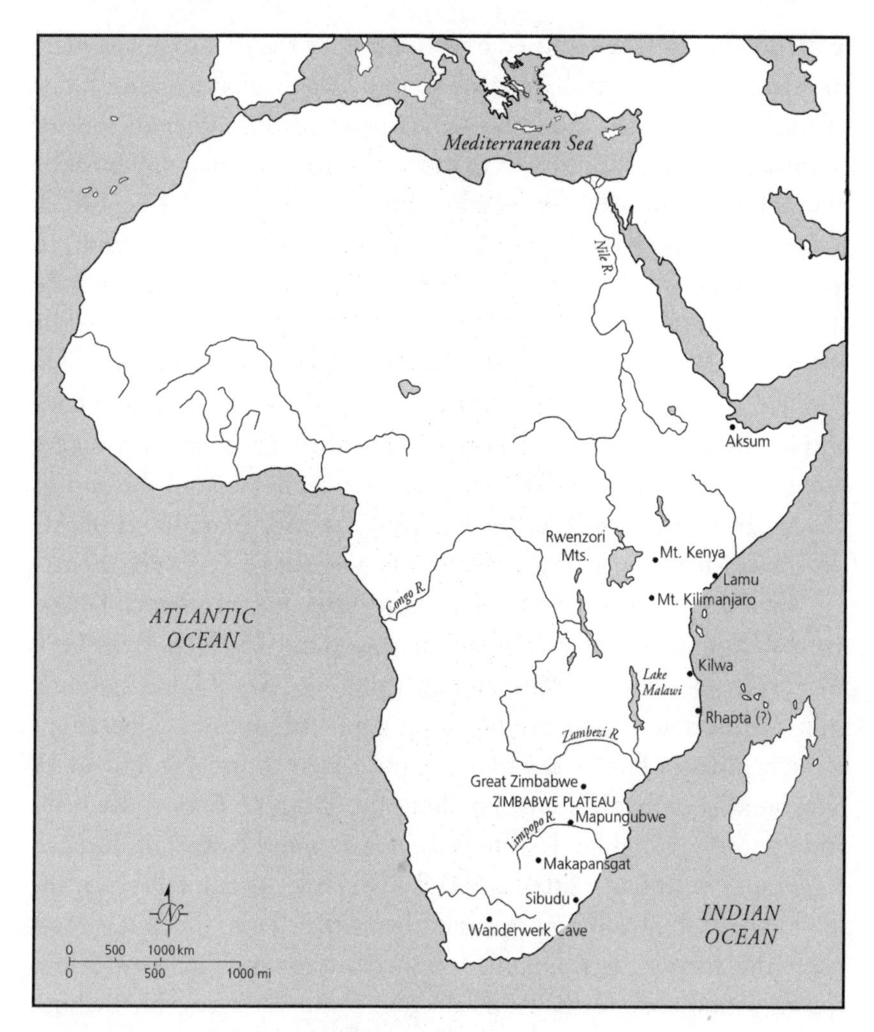

Archeological sites in Chapter 10.

to develop forecasts for the rain-bearing southwest monsoon, using networks of observations throughout India. Enter Gilbert Walker in 1903, a British statistician who used thousands of observations from around the world to establish relationships between the complex atmosphere and other conditions that could affect monsoon rainfall. It was Walker who discovered the Southern Oscillation and its relationship to monsoon rains—a fundamental element in Indian Ocean climate (see Prolegomenon).

Merchant sailors have traveled Indian Ocean waters, from the Arabian Peninsula through Mesopotamia and over to India, for at least 5,000 years. They learned how to navigate the monsoon and were therefore able to control sea trade routes. Deeply guarded knowledge of the Indian Ocean monsoon winds passed from father to son for centuries. The word leaked out to the wider world when a shipwrecked mariner from the Red Sea reached Alexandria and helped a Greek named Eudoxus of Cyzicus make two journeys to India in around 118 to 116 BCE. Soon afterward, Hippalus, another Greek/Alexandrine skipper, worked out a much faster strategy than coasting, using the boisterous August southwesterly monsoon to develop nonstop ocean passages from the island of Socotra off the Red Sea to India and back within twelve months. This dramatic breakthrough in ocean voyaging was to touch dozens of African societies far inland from the Indian Ocean. In this way, global weather patterns affected millions of African subsistence farmers and the chiefs who strove to govern them.

The Red Sea coast of Africa had long attracted merchants. Twelve ancient Egyptian sources ranging from around 2500 BCE to 1170 BCE refer to the wonderous land of "Punt," or "God's Land," lauding it for its precious resources, including gold and incense. Generations of archeologists have tried to work out where Punt was, but in all likelihood it probably stretched along the Horn of Africa's Red Sea and up into the highlands of today's Ethiopia and Eritrea. Indeed, an Egyptian inscription, dated to 600 BCE, speaks of rain falling on the mountain of Punt and how this rain then drained into the Nile—most likely the arm of the Blue Nile we encountered in Chapter 4. All of the Egyptian texts further indicate that Punt was accessible by both land and sea, which implies that people knew how to sail the monsoons along the Red Sea since at least the third millennium BCE.

Nonetheless, the route was clearly not heavily exploited. Punt and its location were shrouded in mystery and regarded as being of such magnificent importance that (in around 1471–1472 BCE) the formidable Queen Hatshepsut adorned the walls of her Deir el-Bahari temple in Upper Egypt with endless goods from Punt, including images of her slaves carrying its incense trees, baboons, giraffes, cattle, dogs, donkeys, doum palms, and fat-bottomed ladies. The walls also list Punt's many prized resources, such as myrrh, ebony, ivory, and gold. We can assume, given the attention she paid to

Punt, that this was perhaps a pioneering state mission that Queen Hatshepsut wanted to leave as her legacy.

Toward the end of the first millennium BCE, things had moved on somewhat. Merchants were using the Red Sea more frequently, although from classical authors like Strabo and Agartharchides, we learn that it was still a journey of some difficulty, with rocky stretches, no anchorages, and violent surf. Writing in the second century BCE, Agartharchides got a bit imaginative at times, describing how a river ran through the land bringing abundant gold dust, as well as mines with nuggets of gold further south. The route was still a secret, we assume. A century later, the route was far better known, even by outsiders, who could rely on quite widely available sailing directions. The *Periplus of the Erythraean Sea* ("Circumnavigation of the Red Sea") of the first century CE is the best known. Its anonymous author, likely a mariner who knew the area well, described in unvarnished Greek the more southerly African coastline, which was then called Azania and stretched far to the south.[5]

The *Periplus* speaks of plentiful sheltered anchorages and places like Rhapta far to the south (a still undiscovered location), where ivory and tortoise shell abounded. Thanks to the predictable monsoon winds, sailing vessels could cross from the Red Sea and back or from East Africa to the western Indian coast and return within twelve months. The arrivals and departures of the trade wind ships were the major events of the year in sheltered anchorages like Lamu in northern Kenya. Here they offloaded their cargoes into small vessels that traded with remote coastal communities far from the spotlight of history. These were the places where valuable commodities could be obtained to trade in Arabia and India: soft, easily carved African elephant ivory, gold and copper for ornaments, and easily malleable iron ore. There were more prosaic commodities as well, including hut poles from African mangrove swamps much valued for housing in treeless Arabia.

Azania was a languorous world of small villages for centuries, visited only occasionally by merchants from the Red Sea. Everything changed during the tenth century, when demand from the Mediterranean world for gold, ivory, and transparent quartz skyrocketed. This was when Islam took hold, as merchant communities developed close to sheltered bays. Some coastal enclaves numbered in the thousands, dwelling in "stone towns," where powerful merchant families thrived as far south as Kilwa in modern-day Tanzania.

Today, this is known as the Swahili Corridor, a narrow coastal strip where indigenous merchant towns developed close to secure anchorages.[6] In the late first millennium CE, Islam brought political and economic connections with a much wider world and fostered ideological links with those further afield. However, the communities and powerful merchant families in the stone towns of this part of Africa also focused on more local ties. They developed carefully preserved political and social links with the webs of trade routes that extended hundreds of kilometers inland. Small parties carried grain, hides, and seashells, along with salt, much valued by farmers, into the far interior. They also carried other exotica: Chinese porcelain, Indian textiles, and glass beads. The shells and trinkets were basically cheap baubles, worth a fraction of the value of the gold and ivory carried down to the coast. Far inland, seashells were prestigious objects, not so much the commonplace cowrie shells, used as hair ornaments, but more esoteric mollusks. The Indian Ocean *conus* shell became a major symbol of chiefly prestige. Five such shells would purchase an elephant tusk in central Africa as recently as 1856. Gold was the hardest to acquire, for the sources were far to the south on the Zimbabwe plateau. Yet one estimate calculates that at least 567 metric tons of gold passed to the coast over eight centuries, for African gold was a major factor in the global economy of the day.[7]

This informal trade had gone on for centuries. Such exotica as Chinese porcelain vessels, fine and coarser cotton textiles, and tens of thousands of glass beads appear in archeological sites hundreds of kilometers from the ports where they first arrived in Africa. Fortunately for archeologists, many of them can be dated from their designs, while spectrographic trace elements often reveal their points of origin.

The monsoon winds connected the East African stone towns with distant lands and drew them into the global world of long-distance trade. Wind strengths may have varied, and the amount of rainfall climbed or tumbled from year to year, but Indian Ocean commerce endured for many centuries, even after European settlement. One can argue, rightly, that global climate played a role in the history of East and Southern Africa over the past 2,000 years. But what effect did the vagaries of the monsoon have on people living in the distant interior? The answers lie in the complex histories of cattle kingdoms and farming villages high on the Zimbabwe plateau between the Zambezi and Limpopo Rivers.

EXPLORING THE INTERIOR (FIRST CENTURY TO C. 1250 CE)

As we travel inland, we enter a world that was virtually unknown to Europeans until missionary explorer David Livingstone traveled through much of central Africa in the mid-nineteenth century. Patchy Portuguese chronicles and the writings of Victorian explorers describe sixteenth- to nineteenth-century happenings. But almost nothing was known of earlier African history until serious research began during the 1960s, except for spurious claims that Great Zimbabwe was a Phoenician palace. We are entering little explored historical territory compared with many other regions of the world, one with complex climatic dynamics into the bargain.

Our old friend the Intertropical Convergence Zone (ITCZ) moves back and forth between the Northern and Southern Hemispheres in the Indian Ocean. It always remains near the equator but is at its northerly limit at about fifteen degrees north. In January it is at about five degrees south. The ITCZ is a zone of intense rain-cloud development. These movements bring rainfall to Southern Africa during the winter, from November to February. Longer term changes in the position of the ITCZ can result in prolonged droughts. This is but part of a complex meteorological scenario, with both El Niños and La Niñas playing significant roles in drought and excess flooding. When examining the role of climate change in East and Southern Africa, one is juggling with an intricate, unpredictable yo-yo.

Some 2,000 years ago, small groups of farmers and animal herders moved into the Zambezi valley and from there into Southern Africa. They settled in small villages over wide areas of the open savanna, preferring locations without tsetse flies, fatal to cattle, where the soils were relatively light and easily cultivated with simple, iron-bladed hoes.[8]

The newcomers moved into landscapes where sparse numbers of San hunters and gatherers had lived for thousands of years. As the centuries passed, the farming population rose, and the San either adopted the new economies or moved onto marginal lands. While the ancestors of the San were thin on the ground and highly mobile, the farmers were tied to the land, where they cultivated sorghum and two forms of millet, this being long before the introduction of maize from the Americas.

A rugged escarpment separates the interior of Southern Africa between the Zambezi and Limpopo Rivers from the Indian Ocean plain to the east. Hot, lower-lying river valleys dissect the higher ground of the Zimbabwe plateau, which averages over 1,000 meters above sea level. Savanna woodland covered the rolling plains, where there were patches of fertile soils for millet and sorghum and a relatively cool, comparatively well-watered environment.[9] These crops grew in the southern summer, but they required some 350 millimeters of rainfall, at least 3 millimeters of water daily. These realities meant that the fields had to receive minimum annual rainfall of about 500 millimeters while the temperature could not drop below 15°C. The requirements varied from landscape to landscape, but these were demanding crops in an environment with a prolonged dry season and unpredictable rainfall. Extensive grasslands provided excellent grazing for cattle, sheep, and goats, but with a constant backdrop of drought and unpredictable rainfall. This was not an exceptional region for farming, for drought, moisture stress, long dry seasons, and sometimes excessive rainfall and flooding varied greatly from year to year.

Rainfall in Southern Africa falls on a striking east-to-west gradient, with the southeastern part of Southern Africa receiving some 66 percent of its annual rainfall during the southern winter. The ITCZ migrates southward in the Indian Ocean, and an easterly flow from the ocean brings often unpredictable rain to a wide area. These relatively wet climatic conditions played a key role in the extent of farming and herding over the past 1,000 years. Most of it was limited to the savanna woodlands, open savanna, and grassland areas between the Zambezi in the north and the Great Kei River in the south.

An array of lake cores, speleothems, and tree rings reveal significant variations both in rainfall and temperature over the past millennium.[10] To generalize: highly variable medieval warming occurred over a wide area of Southern Africa from before 1000 until just after 1300 CE. A significant peak at about 1250 was one of the warmest of the past 6,000 years, with temperatures up to 3°C to 4°C higher than annual maximum temperatures between 1961 and 1990. Then temperatures cooled, as shown by ocean cores and the changing levels of Lake Malawi inland. The coldest temperatures were from about 1650 to 1850. The coolest temperatures of all occurred around 1700, with another cooler snap about a century earlier. Interestingly, isotope records from mollusks excavated in coastal archeological sites in the Western Cape province and elsewhere chronicle declines in sea surface temperatures at the time. Temperature minima and oxygen isotope records from caves at Makapansgat in northern South Africa correspond with the Maunder Minimum between about 1645 and 1715, which brought intense cold to Europe, while there are also signs of the cold Spörer Minimum (1460–1550). These are, of course, generalizations, as speleothems recorded numerous regional variations. After 1760, warmer conditions slowly returned. But, whatever the details, the rainfall and temperature variations were a constant and major challenge for both village farmers and the states that rose, then dissolved, during the Little Ice Age.

THE REALITIES OF SUBSISTENCE FARMING

After a glance at the wider climatic time frame, our focus now narrows to the past 2,000 years, the centuries since subsistence farmers first settled in south-central Africa. Almost all farming revolved around the village and depended on slash-and-burn (swidden) agriculture. In September, at the end of the dry season, the villagers would clear hitherto uncleared woodland and set fire to their plots. Then they would spread the ash and hoe it into the soil so that everything was ready for the impending rains. Then the waiting began as the daily temperatures rose steadily. Sometimes a few showers would fall on different parts of the landscape, on one village's land and not its neighbor's. Should one plant or not? If you planted, you watched the sky for rain clouds.

Sometimes ample rainfall followed, and the crops grew nicely. But all too often, the crops would wither in the fields. Within weeks there would be hunger, and by spring, starvation. Most farming villages could survive a year's drought using stored grain, but not multiple dry years. The people would fall back on wild plants, game, and their herds—if they had livestock.

Generations of hard-won experience lay behind successful farming.[11]

Managing risk required familiar strategies, which varied from village to village, as did the diversity of agriculture. Well-developed coping mechanisms were created at both household and village levels. These included careful longer-term grain storage, food sharing among fellow kin, and ancient doctrines of reciprocity, which ensured that as few people as possible went hungry. Cooperative labor to clear fields and carry out other vital tasks was routine.

These societies relied heavily on fellow kin and powerful ritual ties to ancestors—the guardians of the land since time began. Rainmaking and rain rituals were powerful catalysts in tribal society, as were powerful bonds with kin living in nearby and more distant communities. Long-established ties of reciprocity, of mutual obligation to provide help, food, and even grain for planting, provided powerful weapons of resilience in times of abrupt climatic shifts and prolonged drought.

Dispersed farming villages with small herds survived through the centuries, especially when the villagers could fall back on sporadic long-distance trade in ivory and other commodities exchanged for grain. Most important of all were the rituals and ancestral ties that originally maintained ties between villages near and far. After many generations, these ancestral links and prized supernatural powers that enabled communication with ancestors transformed what had once been small-scale village societies into an ever-changing political landscape of small-scale chiefdoms. Differences in social status depended on perceived supernatural gifts, membership in what became chiefly lineages, and personal charisma, for here, as elsewhere in the ancient world, power often depended on one's ability to command and retain the loyalty of followers—often fellow kin. Ceremonial offices, well-timed gifts, and reciprocal gestures were the currency of loyalty, as was wealth. This wealth was above all on the hoof, especially in cattle.

Cattle were far more than just a source of meat and milk. These prestigious beasts were powerful indicators of affluence and social ranking, as well as lucrative bride wealth. Surplus male beasts were invaluable currency, large herds an unspoken symbol of political power. But herds large and small were vulnerable to unpredictable rainfall and drought, for their members required water at least every twenty-four hours, in addition to good grazing grass. The chiefs of the plateau and river valleys spent much energy acquiring grain and cattle that cemented their political power.

By the tenth century, plateau society was changing but still depended, ultimately, on a tried-and-true adaptation to an unpredictable environment based on dispersed village communities. Ultimately, success in the plateau environment depended on stock raising and subsistence agriculture as it always had. But the plateau offered far more than fertile soils and grazing grass. Gold extracted from alluvial deposits and quartz reefs, as well as copper, iron, and tin, soon became staples of long-distance trade. Not only cattle but ivory-tusked elephants flourished on the plateau. The Shona chiefs who controlled the highlands came in contact with visitors in search of gold and ivory from the distant Indian Ocean coast. At first, the coastal trade was sporadic, but it expanded dramatically after the tenth century, during a period of warmer and damper climatic conditions. Inevitably, some highland chiefs acquired economic and political dominance, as they managed to control trade routes and extract tribute from lesser chieftains. This gave them volatile political and economic power that could evaporate rapidly in the face of prolonged drought or changes in the Indian Ocean trade.

MAPUNGUBWE AND GREAT ZIMBABWE (1220 TO C. 1450 CE)

A powerful kingdom developed in the hot, low-lying Limpopo River valley.[12] Between about 1000 and 1300 CE, during the centuries of the Medieval Climate Anomaly (MCA), higher rainfall caused regular flooding and turned the normally arid valley into a landscape where not only farming but also cattle did well. The good Limpopo grazing grass also attracted large elephant herds. All the elements for a commanding kingdom were in place, especially when circumstances favored a society in which cattle were symbols of wealth and social ranking. At first, the newly powerful chiefs dwelt in larger villages in the valley. By 1220 CE, a small group of spiritually powerful individuals moved atop a conspicuous hill named Mapungubwe, overlooking the valley. The distinctive hill had long been a center for rainmaking rituals, a major element in local Shona society. The more abundant rainfall of the time seems to have given Mapungubwe's leaders powerful spiritual validity.

Mapungubwe, with its great wealth in cattle, gold, and ivory, was not alone. Other centers emerged, many of them located on flat-topped hills with stone-walled, safe cattle corrals, each carefully located in major river basins to ensure reliable water supplies. Chiefs and villagers adopted deliberate strategies to adapt to seasonal changes and climatic shifts. They selected farming soils with care, planted drought-resistant crops like sorghum, and paid careful attention to grain storage for dry years. Their farming strategies flourished over a wide area, including attempts to reduce cattle mortality from tsetse fly bites by lighting smoky fires.

As rainfall declined during the thirteenth century, the ability of the Mapungubwe chiefs to intercede with the supernatural forces that controlled rain assumed ever-greater importance. Rainmaking rituals became more centralized and increased the legitimacy and power of the chief. But he became more vulnerable as dry season followed dry season and his powers as a rainmaker appeared significantly diminished. Between about 1290 and 1310 CE, cooler temperatures and increasing aridity, combined with highly variable rainfall, slowly undermined the credibility of the chief and his ability to guarantee the fertility of the Limpopo floodplain. Mapungubwe's influence declined broadly during more prolonged thirteenth-century droughts.

Mapungubwe was not alone. The rich biodiversity of the entire region supported numerous communities that engaged in local and long-distance trade in basic commodities, metals, and imports like Indian Ocean beads, seashells, and textiles. A great deal of the long-term insurance against climatic shifts came from cooperation with close neighbors and people living further afield. Much also depended on making use of others' skills. Among them were copper and iron working, mining raw ores, and even making iron spears for elephant hunting. Some of these societies acquired considerable social complexity, but, above all, they were adept at risk management through both agricultural expertise and craft production, as well as sharing knowledge and experience among different communities and kin groups. We still know almost nothing about these many communities in the Mapungubwe region or their vulnerabilities.

Without question, there was a connection between the greater prevalence of drought and the decline of chiefly influence as Mapungubwe's ability to adapt to drier conditions was undermined. During the early fourteenth century, political power shifted northward from the Limpopo valley to the Zimbabwe plateau. Great Zimbabwe, with its iconic stone structures and overlooking hill, is world famous.[13] What is not so well known is that the site began, and indeed continued, as a major center of powerful rainmaking rituals. Zimbabwe lies at the edge of a gold-mining area, but, more importantly, the area remains fairly green throughout the year, thanks to the frequent mists and rain that move northward from the nearby Mtelikwe valley, straight from the Indian Ocean. This isolated, seeming oasis in a comparatively dry area of the plateau was revered as a rainmaking center. The imposing hill with its mighty boulders and caves became a focal point for rainmaking and ancestor rituals associated with the Mwari cult, which played an important role in Shona society. The chief medium of the Mwari, a monotheistic belief, wielded significant influence in Zimbabwe society.

Access to the hill must have been restricted, but Zimbabwe's chiefs presided over an extensive kingdom supported by subsistence agriculture and by cattle, as at Mapungubwe a source of wealth and a basis for social ranking in communities large and small. But cattle are a demanding source of wealth, thanks not only to disease but to their need for extensive grazing grounds and, above all, ample water supplies. Growing population densities, overexploitation of woodland for firewood and other purposes, and, above all, cooler and drier climatic conditions chipped away at the resilience of a kingdom in an area of unpredictable rainfall and often soils of only moderate fertility. For all their efforts to store grain and maybe centralize food supplies, Great Zimbabwe's chiefs had few long-term protections against climatic

Aerial view of Great Zimbabwe with the Great Enclosure in the foreground and the imposing Hill Ruin in the background. *Christopher Scott/ Alamy Stock Photo.*

stress, except for one—the prestige, wealth, and power brought by the Indian Ocean trade.

The political situation surrounding the Zimbabwe kingdom and its successors was complicated. Succession was a complex issue in a kingdom where the royal cattle herds were too large to be kept at the capital. This meant that numerous other polities surrounded it—many of them with their own spirit mediums—presenting alternative power centers. Chiefly capitals moved frequently, many of them marked today by smaller-scale stone buildings. Warfare was apparently commonplace, although limited, given the need for people to cultivate and harvest the land.

Cattle kingdoms like Great Zimbabwe may have been ruled by powerful chiefs, who were the beneficiaries of the gold and ivory trade, but their dominance relied not only on the prestige conferred by the Indian Ocean trade but also on their herds, which required wide areas of grazing land. The main centers may have controlled long-distance trade, but they were much more vulnerable to climatic shifts than the dispersed villages that formed most of their domains. Populations around larger centers were denser, needing reliable food supplies, whereas small villages could rely on hunting and foraging and shift more readily to alternative crops more resistant to arid years. The main stress was not the frequent arid years but longer-term droughts, which could decimate not only grazing but permanent water sources. Cattle diseases like rinderpest, locust plagues, and even occasional floods were added hazards. Under these circumstances, large, widely dispersed herds provided some protection against hunger, as did wild plants and carefully organized hunting.

Why exactly the Zimbabwe kingdom disintegrated remains a mystery, but most likely a concatenation of events came into play. One may have been the exhaustion of gold workings close to Great Zimbabwe, which would have caused traders to search elsewhere. The early fifteenth century saw a return of cooler and drier weather conditions that may have undermined the agriculture that supported the growing population. More intense competition from neighboring kingdoms may also have played a role. By the 1560s, the vulnerabilities of larger kingdoms like Zimbabwe were much greater than those of more dispersed tribal groups. The center of political gravity had now moved northward toward the Zambezi River region, by which time Portuguese traders and settlers had penetrated deep onto the plateau in search of gold. Between 1625 and 1684, the Portuguese wrested control of mine workings from local chiefs, undermining the economic prosperity of once powerful kingdoms. Nevertheless, traditional food systems, rainmaking rituals, and the survival of many larger communities persisted in the face of political instability.[14]

No Southern African kingdom, even Zimbabwe, was long-lived or of major size. There was none of the large-scale infrastructure that bound together the centralized Khmer empire. This was a world of dispersed villages, whose resilience depended on careful risk management, and volatile kingdoms, which rarely lasted more than two centuries, ruled by chiefs whose loyalty depended on generosity and the polygamy that reinforced it. Anyone who ruled a kingdom on the Zimbabwe plateau had to be an entrepreneur and a politician. Much of chiefs' security depended on their skills with people and their ability to acquire cattle. Here our knowledge still falters, except for a certainty that every kingdom in the interior was fragile. What ultimately survived was the dispersed village with its reservoir of knowledge about managing risk in a very challenging environment. This remains true in modern Africa, where subsistence agriculture and the village still support millions, despite rapid urbanization. Coping with climate change was, and still is, most effective at the local level, where people know the environment and the landscape.

This chapter links long-distance, global trade with the world of the village farmer, people with very different perspectives from those of a Khmer rice farmer or a European feudal peasant. The warming centuries associated with the Medieval Climate Anomaly had important consequences in much of tropical Africa. In the next chapters, we explore both the MCA and the Little Ice Age in Europe and North America.

11

A WARM SNAP

(536 to 1216 CE)

By any standards 536 CE was a perfectly horrible year across the eastern Mediterranean (see also Chapter 5). The Byzantine historian Procopius wrote, "The sun gave forth its light without brightness, like the moon, during the whole year."[1] Eighteen months of what appeared to be fog cast darkness over Europe, the Middle East, and parts of Asia. The culprit was a massive volcanic eruption in Iceland, which threw enormous quantities of ash across the Northern Hemisphere. Two more huge eruptions followed in 540 and 547. A combination of these volcanic events and the Justinian Plague plunged Europe into economic stagnation for over a hundred long years, until 640.

VOLCANIC TURBULENCE (750 TO 950 CE)

Volcanic eruptions cast sulfur, bismuth, and other substances high into the atmosphere. They form an aerosol veil that reflects sunlight back into space and cools the earth. Researchers first identified the 536 CE eruptions as spikes in ice cores from Greenland and Antarctica. Subsequently, a seventy-two-meter core drilled into the Colle Gnifetti glacier in the Swiss Alps during 2013 yielded records of volcanic eruptions, Saharan dust storms, and human activities obtained from laser-carved ice slivers that represented a few days or weeks of snowfall.[2] About 50,000 samples per meter of ice allowed glaciologist Paul

Mayewski and his colleagues to pinpoint events like volcanic eruptions—even lead pollution—down to the month going back 2,000 years. In the case of the 536 CE volcanic eruption, the ice particles identified the source as Iceland.

On a global scale, volcanic activity was never continuous. Despite the events of 536, there are few other signs of volcanic activity during the first half of the first millennium CE. But the two centuries between 750 and 950 were another matter. The world experienced at least eight massive volcanic eruptions. We know all this thanks to data from the Greenland Ice Sheet Project 2 (GISP 2). The GISP 2 ice core provides a notable record of atmospheric chemistry that reveals information on weather events in Siberia, storms in central Asia, and ocean tempests. Evidence of volcanic eruptions appears in the Greenland ice cores as sudden surges of background levels of sulfate particles. Most of their origins are still unknown. Remarkably, the chronology obtained from the GISP 2 core for 750 to 950 is accurate to within two and a half years and more precise for the main eight eruptions of the period.

But how does the science compare with contemporary written documents? Historians Michael McCormick and Paul Dutton worked with glaciologist Paul Mayewski to compare the most important volcanic events between 750 and 950 with surviving historical accounts. They used only winters recorded as exceptionally severe in several regions—not just local observations. A combination of glacial cores and firsthand accounts from people alive at the time produced a dizzying story of severe winters and occasional wet summers with crop failures and famine. Eight of the nine harshest winters in western Europe between 750 and 950 show links between spikes in sulphate deposit levels in the GISP 2 ice core and historical records complaining of unusual cold. The winter of 763–764 brought great suffering from Ireland to the Black Sea. Irish records refer to snowfall that lasted almost three months. The bitter cold spread across central Europe as Constantinople suffered from cold so intense that ice extended 157 kilometers out from the north Black Sea coast. When the ice melted in February, ice floes blocked the Bosporus.

Very harsh winters returned in 821–822 and 823–824, following two wet summers with poor wine harvests in the Frankish kingdom, which then covered most of western Europe. The Rhine, Danube, Elbe, and Seine froze to the point that carts could cross the great rivers for thirty days or more. In 855–856 and 859–860, the cold struck again. The winter of 859–860 was exceptionally long and cold throughout western Europe. In Rouen, the freeze began on November 30 and endured until April 5. The late summer of 873 brought a plague of locusts to Germany, Gaul, and Spain, followed by a bitter winter. Famine and its related diseases killed about a third of the population of Gaul and Germany. The years from 913 to 939–940 were also very cold, the latter attributed to the Eldgjá volcano in Iceland.

Volcanic eruptions can have a significant impact on climate. Major eruptions can inject vast amounts of volcanic gas, ash, and other substances into the stratosphere. Volcanic gasses such as sulfur dioxide can cause global cooling. When sulfur dioxide converts into sulfuric acid, the latter condenses rapidly in the stratosphere to form sulfate aerosols. These increase the reflection of radiation from the sun back into space, which cools the earth's lower atmosphere. The huge eruption of Mount Pinatubo in the Philippines in June 1991 injected some twenty million tons of sulfur dioxide into the atmosphere at an altitude of thirty-two kilometers. The disturbance cooled the earth's surface by more than 1°C at its height. Much larger eruptions in the past, like those of Tambora and Krakatoa in the nineteenth century, cooled temperatures for several years. No one claims that the seeming frequency of volcanic events during the late first millennium CE collapsed entire kingdoms, but they had a powerful impact on the climate of these two centuries, which affected crop yields and both animals and humans. The demographic losses were severe in the harsh years, as were the economic reverses in food supplies. On a larger stage, King Charlemagne of the Franks (742–814) was relatively lucky in his experience of rapid climate change as his people lived through the terrible winter of 763–764 and the famine of 792–793. However, his worried son, Louis the Pious (778–840), was so obsessed with a seeming link between yet another terrible winter of 821–822 and God's wrath that he paid public penance for both his and his father's sins in August 822. To no avail: a year later another vicious winter froze his empire.

THE MEDIEVAL CLIMATE ANOMALY (C. 900 TO 1200 CE)

Just over four centuries later, in 1244 CE, a Franciscan friar, Bartholomew the Englishman, declared that Europe comprised a third of the known world, from the "Northern Ocean" to southern Spain.[3] Scholars of the day gazed over a broad landmass. The easterly reaches ended in the seemingly endless European plain that melted over the distant horizon into the Asian steppe. There, the population was sparse, mostly nomadic bands constantly on the go, their movements driven by irregular cycles of drought and more plentiful rainfall. The semiarid grasslands breathed like giant lungs, sucking in animals and people when rains came, expelling them to better-watered landscapes on the margins during drier times. Hardly surprisingly, medieval Europeans thought of themselves as surrounded by a dangerous human and natural world. Islam pressed upon them from the east; the Atlantic formed a barrier in the west. Nomadic tribespeople from the eastern plains hovered on the fringes of Eurasia.

The eastern steppes were the land of Genghis Khan, who once called himself "the Flail of God." The threat was real. Fourteen years after Genghis Khan's death in 1227 CE, a Mongol army defeated the princes at Legnica, now in Poland. Nine sacks containing the right ears of slain Poles arrived at the Mongol court. The invaders withdrew abruptly to the east in 1242 for reasons that remain a mystery. It may be no coincidence that heavy rainfall and cooler temperatures restricted Mongol cavalry and reduced forage for their horses.[4]

One cannot blame the Mongols for casting covetous eyes on the better-watered, fertile lands to the west. Europeans lived on a peninsula hedged in by more arid environments. Between about 1000 and 1300 CE, climatic conditions were somewhat warmer, with slightly higher temperatures than in earlier years. These three centuries, commonly known as the Medieval Climate Anomaly, briefly turned Europe into a prosperous granary.

We owe the notion of an unusually warm medieval period to perceptive English climatologist Hubert Lamb, who in 1965 coined the name Medieval Warm Epoch—though it was later renamed the Medieval Warm Period and is today known as the Medieval Climate Anomaly (MCA).[5] Unlike modern climatologists, Lamb had few climatic proxies to work with and depended mainly on a patchwork of historical sources. He was one of the first scientists to suggest that climate could change within generations, opposing the then orthodox view that climate was a long-term constant. Lamb noted modest but persistent shifts in winter circulation over the North Atlantic and Europe during medieval times. He also observed, "In England particularly it seems that there must have been less liability to frost in May in the period 1100 and 1300 AD." This augured well for harvests.

The climatic blip occurred during the High Middle Ages (1000 to 1299 CE). Lamb connected it to art historian Kenneth Clark's notion of "the first great awakening in European civilization," as immortalized in Clark's BBC TV series *Civilization*, aired in 1969. Although the MCA is still little understood, especially outside Europe and North America, it has become established climatic canon in many climatological circles. But did this period actually exist as a well-defined entity, with closely delineated boundaries?

Today's fine-grained archeology shows how many rigid classifications of ancient human societies bear little resemblance to the cultural reality of the past. The past was dynamic, ever changing, and rarely comprised of sharp lines in the sand. As such, archeologists regard the artificial subdivisions of the human experience simply as convenient terms of reference—as helpful tools. The same can be said of the MCA. While most climatologists agree that this peculiar period lasted from about 950–1000 to 1250–1300, there are umpteen variations on the limits, many of them fluctuating by locality.[6]

We've already described a wide range of ancient societies that came into being and imploded during the MCA centuries in areas outside Europe, but the European climate pattern is still not well defined elsewhere. During the 1970s, climatologist V. C. LaMarche used tree rings and other data from the White Mountains of California to show that climatic conditions were mostly warmer and drier between 1000 and 1300, then cooler and wetter from 1400 to 1800. These shifts resulted from a north-to-south shift of the storm track over the region. This discovery made it likely that there was a global shift in circulation patterns affecting a far wider area than Europe.

But there's more: studies of the intensity of sunspot activity document five low points over the past 1,200 years. These usually coincide with cooler temperatures. The earliest lasted from 1040 to 1080 CE, which coincides with the MCA. Then came a cluster of four more lows in sunspot movement after 1280, described in Chapter 12, which coincides with the Little Ice Age of the late sixteenth to early seventeenth centuries.

The kinds of climatic perturbations that occurred over these centuries were predominantly local, even if they originated in larger-scale interactions between the atmosphere and the ocean. Europe enjoyed lengthy warmer cycles, with temperatures slightly cooler than today. The eastern Pacific was cool and dry with La Niña conditions, causing megadroughts in the North American Southwest (see Chapter 8). The western Pacific and Indian Ocean were warmer; the North Atlantic Oscillation trended toward a positive phase that brought warmer temperatures and greater storminess; the Arctic supported much less summer ice. There's sketchy evidence for somewhat warmer temperatures from Tibet to the Andes and in tropical Africa. In brief, temperatures between 1000 to 1349 were warmer than those experienced between about 1350 and 1899 on all six continents. However, temperatures from 1900 to the present have been warmer across the world, with the exception of Antarctica, where the surrounding oceans may cause climate inertia. No one yet knows.

Rich proxy archives tell the story of Europe's late medieval climate. They reveal a period of constant, sometimes quite dramatic temperature shifts. For instance, climatologists Ulf Büntgen and Lena Hellman have compared climatic data from both glacial cores and tree rings in the European Alps.[7] They show relatively warm medieval and recent temperatures, with a colder interval between them. Sections of high-altitude larch trees from the western Swiss Alps were used for tree ring dating, coming both from trees themselves and the wood of historic buildings from the era. High temperatures in the tenth and thirteenth centuries were like those of modern times. There has been considerable natural variation over the past millennium with, however, relatively warm conditions before about 1250 and after about 1850. Six of the warmest decades between 755 and 2004 occurred during the twentieth century. The coldest year was 1816, the warmest 2003, with a variation of 3.1°C between the warmest decade in the 1940s and the coldest in the 1810s. The climatic clues come from sometimes unexpected sources. Even tiny midges recovered from a glacial lake in the Swiss Alps can be used as climatic proxies, providing estimated July temperatures as far back as 1032. The midges tell us that medieval temperatures were about 1°C warmer than those from 1961 to 1990.

Another impressive body of research using oak tree rings from central Europe focuses on April to June rainfall shifts, using thousands of rainfall-sensitive tree ring series from a large area covering the past 2,500 years.[8] The research involves far more than tree rings and includes comparisons to instrument records of the time along with historical records. No fewer than eighty-eight eyewitness accounts of rainfall conditions coincided with thirty out of thirty-two extremes in the tree ring record—an impressive correlation. By combining the oak record with those from other trees, like high-altitude larches from the Austrian Alps, the researchers produced a composite record that correlates remarkably well with the variations in June–August temperatures recorded by modern meteorologists between 1864 and 2003.

By the early 800s, extreme climate conditions caused by the rash of volcanism calmed and more closely resembled those of Roman times, albeit still with much volcanic activity and some very harsh winters. This was when new European kingdoms of the High Middle Ages took shape. Between about 800 and 1000, they began their dramatic cultural and political achievements.

SUBSISTENCE AND DRUDGERY (1000 CE)

In 1000 CE, Europeans depended almost entirely on subsistence agriculture. In that respect, they were the same as Angkoreans, the Maya, or the Ancestral Pueblo. Farming practices were still remarkably

Locations in Chapters 11 and 14.

simple and exceptionally vulnerable to sudden climatic shifts or to the often serious environmental effects of volcanic episodes.

Europe's rural landscape was a tapestry of forest and woodland, of river valleys and wetlands, much modified by thousands of years of farming and herding.[9] By 1000, most rural people lived in either scattered hamlets or, more commonly, in larger villages, surrounded by open fields divided into long strips of about 0.2 hectares apiece. Wresting a living from Europe's soils was never easy, but it could be done, especially during cycles of warm, relatively dry summers, with temperatures in May warm enough for planting.

Subsistence farmers in England and France grew mainly wheat, barley, and oats. About a third of arable land supported wheat; perhaps half was planted with barley and the rest with various crops including peas. By modern standards, the yields were tiny, about a fourth of today's. Of the four hectoliters per 0.4 hectares harvested, 20 percent went back into the soil as seed for the next crop. Add church tithes and the tax in grain paid to the landowning lord even in poor harvest years, and there was precious little left to support a family. A farmer with a wife and two children could barely survive on the crops from two hectares, with little margin for unexpected frosts, droughts, or storms. Everyone, even children, grew vegetables and foraged for mushrooms, nuts, and other wild plant foods.

Most families had some livestock, perhaps a couple of milk cows, pigs, sheep, or goats, as well as chickens. If they were lucky, they owned a horse or at least had access to one for plowing. Animals provided meat and milk, in addition to hides and wool. Every part of the animal was used. Most of the year, the beasts roamed free, but it was a constant battle in winter to keep breeding stock alive and fed. Every autumn, villagers slaughtered surplus males and old cows to allow their precious hay stocks to last longer. Each June and July, people mowed their hay, praying for good weather, so that the crop could be stored dry and not rot or burst into flames from spontaneous combustion (a problem still experienced by farmers if they store bulky wet crops rapidly). A wet year had grievous consequences, with potentially catastrophic stock losses.

The endless cycle of winter, spring, summer, and autumn defined the subsistence farmer's life and mirrored the verities of human existence. Planting, growth, and harvest, then the quiet months: cycles like human existence, a matter of birth, life, and death. Life was unremittingly harsh. Hunger was a given. Everyone in a medieval farming village experienced malnutrition at some point, sometimes famine, starvation, and subsequent famine-related diseases. Infant mortality rates were astronomic; life expectancy for most peasants was in the twenties.

Like the peoples of ancient civilizations before them, medieval farmers knew the surrounding environment like the backs of their hands. They were familiar with the properties of different grasses; they understood that exhausted soil could be cultivated once more and that arable land had to be grazed and manured by animals, then rested to regain its fertility and minimize plant diseases. Everyone knew the seasons for every kind of edible or medicinal fruit and plant.

Wheat farming was inefficient, relying as it did on very simple tools and brutally hard work. Survival depended on knowledge acquired through experience in the fields, passed down through generations. For instance, scattering too little seed left space for weeds to grow, while too much seed choked the seedlings. There were no carefully calibrated formulas, just folklore and practical experience. Like subsistence farmers everywhere, medieval cultivators were experts in risk management, in planting as diverse a range of crops as possible. Rainfall was often more plentiful here than in the tropics, but teasing crops from the land, let alone a food surplus, was just as hard as it was in the Maya lowlands or at Angkor Wat, if not harder.

Historically informed climate change skeptics seize on the Medieval Climate Anomaly as one of continuous, benign summers, as something positive and a symbol of natural warming—claiming it to be just like today's. The argument is founded on the idea that the type

Pieter Breughel, *The Harvesters*, 1565, provides a somewhat misleading impression of medieval European agriculture on a hot summer's day. The reality was much harsher. *Universal Images Group North America LLC/ Alamy Stock Photo.*

of warming we are now experiencing has happened in the past and is therefore not a sign of instability or crisis. This is, of course, nonsense. Not only have we humans caused the current warming, but our growing body of evidence does not support the vision of bygone lovely long summer days of nonstop plenty.

WARMING UP (800 TO 1300 CE)

There were certainly years when farmers basked in summer and crops grew vigorously in bright sunshine. But climate is never static and changes repeatedly, often in inconspicuous ways. Europe's farmers enjoyed slightly warmer and drier conditions, but tree rings chronicle constant minor shifts resulting from still little-known changes in the earth's obliquity, cycles of sunspot activity, and volcanic eruptions—to mention only a few factors. Between about 800 and 1300 CE, the endless dance between the atmosphere and the oceans slowed slightly into a more measured routine during a period when Europe changed profoundly. But everyone still lived from season to season: the long days and warm temperatures of summer, the season of darkness that was winter, lit at best by flickering candles and smoky fires, when people huddled together for warmth. A thick robe or a cozy bed was a great luxury.

Nonetheless, at its best, this warm period was, as the climate skeptics observe, a delightful time to be alive, and Europe certainly prospered during the three centuries of the High Middle Ages (roughly 1000–1250 CE). William of Malmesbury, a monk and historian, traveled through England's Vale of Gloucester in about 1120 and admired the summer landscape. "Here," he wrote, "you may behold highways and publick roads full of fruit-trees, not planted, but growing naturally."[10] He admired English wines, "little inferior to the French in sweetness." Much to the consternation of local vintners, wine from across the channel flooded onto the French market. This is hardly surprising. At the time, vineyards flourished as far north as southern Norway.

Most important of all, the growing season for cereals in warmer years was as much as three weeks longer. Between 1100 and 1300 CE, the May frosts that had plagued earlier farmers were virtually unknown, a welcome prelude to often long, settled summer weather during the growing and harvest seasons. Rural populations grew steadily, as the limits of cereal farming expanded dramatically, often onto land that was considered marginal in earlier times. Kelso Abbey in southern Scotland had over one hundred hectares under cereal cultivation at three hundred meters above sea level, well above today's limits. The monks also owned 1,400 sheep and supported sixteen shepherds' families on their land. Norwegian peasants grew wheat as far north as Trondheim. At lower altitudes to the south, cereal yields increased significantly, thanks to the longer growing season. The resulting food surpluses fed growing towns and cities. Meanwhile the demand for arable land skyrocketed as primordial oak forests fell before the axe. Not that life was easy for most people. A single urban family of moderate means would purchase five and a half tons of grain a year. Most of it was milled into bread. Most families living above the poverty line consumed a 1.8-kilogram loaf of bread a day. The poor often ate frumenty, a porridge of cracked wheat or other grains boiled with milk or broth. But bread and beer were the main dietary staples, yielding about 1,500 and 2,000 calories per person daily.[11]

Yet there were still some intensely frigid winters, like that of 1010–1011, which gripped even the eastern Mediterranean in severe cold. Despite occasional fluctuations, persistent warmer conditions melted ice caps, raised mountain tree lines, and triggered sea level rises of up to eighty centimeters in the North Sea. By 1100, a tidal inlet reached as far inland as the town of Beccles in Norfolk, turning it into a flourishing herring port. The rising sea levels also allowed vicious storm surges from violent westerly gales to inundate low-lying coasts, especially in the Netherlands. Great storms in 1251 and 1287 rushed ashore and formed a huge inland waterway, the Zuider Zee. Moreover, Europe was not at peace despite the sunshine of the MCA. Violence was endemic. Fleeting alliances and brutal military campaigns preoccupied the elite and privileged. Knightly displays of courage and power were ways of assessing the limits of political authority and establishing who exploited them. Warfare ebbed and flowed, made possible by food surpluses and the ability of ambitious lords to build strongholds and castles to protect a growing population.

Eventually, rapid population growth and a growing volume of long-distance trade changed the political landscape as more permanent kingdoms emerged. The distant beginnings of modern Europe first appeared. North of the Alps, more people on the land led to massive clearances of forests and marshes, including areas that had become wilderness after being cultivated during Roman times. People moved onto marginal lands with thinner soils. Thousands of farmers moved eastward across the Elbe River. These were the centuries when the Catholic Church was at the peak of its political power, marked by the Crusades against the Seljuk Turks and Fatimid Egypt, the founding of the Crusader States in the Levant, and the overthrow of Islamic al-Andalus in Spain.

Europe in the High Middle Ages enjoyed a new explosion of art and intellectual discourse that combined ideas from thinkers like Aristotle and Thomas Aquinas with concepts from Islamic and Jewish philosophy. This was when monarchs encouraged the building of Gothic cathedrals; it was also the era of illuminated manuscripts and ethereal woodwork, to mention but a few achievements. All of these innovations, whether intellectual, material, spiritual, or sociopolitical, depended on abundant food surpluses that generated wealth and money to pay for artisans and growing numbers of nonfarmers and to glorify God. When harvests were bountiful and food plentiful, everyone, whether monarch, noble, or commoner, gave thanks and endowed God with lavish gifts. Everyone feared divine wrath, manifested in famine, plague, and war. When harvests were poor, gifts evaporated, and cathedral building slowed. Donations or no donations, the magnificent panoply of medieval Europe depended ultimately on the anonymous labor of subsistence farmers in the countryside that now lapped growing towns and cities.

Monarchs, nobles, the religious, and town and city dwellers depended on cereals provided almost entirely by local farmers. Most people consumed simple diets—bread, dry biscuits, porridge, and soups. They added fresh or preserved fruit or vegetables, and occasionally meat, to what was a monotonous diet. Meat was too expensive for most people to eat frequently; fish, unless pickled or salted cod or herring, was a coastal, lakeside, or riverbank staple. Even a minor crop failure caused grain prices to rise and hunger to descend on rural communities, leaving their inhabitants more vulnerable to disease. In growing cities, social disorder and hunger went hand in hand in the form of bread riots.

For almost 1,000 years, European subsistence agriculture had chugged along. Europe's economic and social system depended on feudal land ownership in the hands of local lords and the church. The peasantry, who cultivated the soil, lived from season to season, using tools virtually unchanged for centuries. Metal implements of tillage were in short supply, so many farmers relied heavily on wooden tools that barely scratched the surface of the soil. Few could afford horses or oxen to pull their plows and relied on family members instead. Effectively, they practiced a simple form of monoculture, which drained the soil of important nutrients and reduced crop yields, even after fallowing. Given the rapacious ecclesiastical and lordly habit of increasing taxes in kind even in good harvest years while the peasant's share remained the same, there were no incentives to increase production. Occasional shortages were one thing, and the system survived. But when significant changes in rainfall and temperatures intervened, the foundations of daily life and farming came apart, as they did beginning in 1314.

DARK DAYS AND THE GREAT FAMINE (1309–1321 CE)

Europe experienced very cold winters between 1309 and 1312 CE. Pack ice extended from Greenland to Iceland, so thick that polar bears could walk across from one to the other. The North Atlantic Oscillation (NAO) had been in high mode, with low pressure centered over Iceland, which accounted for the cold. Then, abruptly, the NAO switched to low and unstable conditions and swooped into Europe. No one knows why, but some atmospheric spasm caused a huge air mass to rise over northern Europe, condense into water, and dump torrential rain over a large area.[12]

Make no mistake: these were epic rains that began in mid-April or May 1314, around Pentecost. The abbot of Saint Vincent, near Laon in northern France, wrote, "It rained most marvelously and for so long."[13] Another account tells us that it rained for 155 days without a break over a wide area from the Pyrenees to the Urals far to the east and as far north as the Baltic. A Salzburg chronicler observed, "There was such an inundation of waters that it seemed as though it was THE FLOOD." Disastrous floods swept away more than 450 villages in Saxony alone. Bridges fell apart; dikes were overwhelmed. Even buildings with sturdy foundations collapsed.

The greatest devastation took hold in the fields. Generations of population growth and ardent deforestation had exposed compacted, thin soils, especially on less productive, marginal lands on slopes and higher ground. The pounding rainfall turned fields into mud, washed away the now-exposed, compacted soil, and formed deep erosion gullies that cratered farmland. By the fourteenth century, plows had dug deep furrows into the best topsoil. Under normal conditions, such plowed fields could absorb the average annual rainfall of 760 millimeters without difficulty. The 1314 deluge brought five times the rainfall, at least 2,540 millimeters. Long-furrowed topsoil washed away, exposing nothing but hard clay subsoils. The lighter, more marginal soils melted away in days. From Scotland and England across northern France and east to Poland, almost half the available arable land vanished, leaving rock behind.

There was hunger, accompanied by a very cold winter. Spring 1315 brought more inexorable rain. As was customary, everyone planted, but after four months of unremitting downpours, there was no harvest in England or northern France. As was the classic strategy across time and civilizations, many European families abandoned their land, wandered aimlessly, or sought charity from relatives. By late 1316, thousands of laborers and peasants were reduced to penury. Communities dissolved or shrank, especially those on marginal farming land, deserting their villages and their fields because of a lack of seed corn and draught oxen. Often there were not enough people to cultivate or plow.

Operating water mills had been a profitable business, tightly controlled by those who owned them and collected taxes in kind for using them. The deluge of 1315 and 1316 swept away hundreds of water mills; others flooded and became unusable. These were the very mills used to grind the people's staple, wheat, but grain was in short supply anyhow. The rains reduced the productivity of the land drastically—not only because they made harvesting and planting hard, often washing away the food, but also because the floods leeched nitrates from the soil. The wet weather brought plant diseases to depleted soils, especially molds and mildews. By 1316, the yields of both wheat and barley crops around Winchester in southern England were a mere 60 percent of the usual average, the lowest between 1271 and 1410. They remained 25 percent lower for at least another five years.[14]

Britain's royalty gave safe conduct to Spanish grain merchants in the hope of increasing grain supplies. But the same weather pattern reappeared in 1316. By now, the cumulative effects of years of excessive rainfall and constant floods were devastating. As we witnessed earlier, droughts in the tropics can wither growing crops and turn farmland into arid wilderness. When the rain returns, however, crops can be grown again in short order, and the drought is forgotten. But extreme rainfall with its floods, as during the years 1315 to 1317, caused lasting damage that took years to mitigate. German chroniclers wrote of fertile, now barren farmland. The anonymous biographer of English king Edward II, writing in 1326, when memories were fresh, remarked, "The floods of rain have rotted almost all the seed, to such an extent that the prophecy of Isaiah might seem now to be fulfilled . . . [I]n many places the hay lay so long under water that it could neither be mown nor gathered."[15] Then the winter of 1317–1318 arrived with savage cold due to an increase in the temperature gradient between the North Atlantic Gulf Stream and the Arctic Ocean.

The rains affected every aspect of the medieval diet. In those days, no one laid down wine to produce vintages. They consumed it within a few months. In 1316, effectively no wine was produced in France with the failure of the grape harvest. Salt, mainly created by using sunshine and burning in coastal salt pans, became rare and expensive, for firewood was too wet to set alight. The prices of salted cod and herring soon rose to their highest levels in a century.

Malnutrition was an obvious consequence of the food shortages. Chronic warfare and brutal despoliation of crops and other food by roaming armies turned rising levels of malnutrition into starvation. The effects could be seen on every side, with millions dying of starvation across Europe. In 1319, the bodies of those who had perished from hunger in Winchester, England's former capital city, littered and stank up the streets. Desperate, some turned to cannibalism; others, to infanticide. As always, the poor and rural peasants suffered the most. The rich and denizens of religious communities usually had an adequate and varied diet. Not that this was always the case. In wealthy Flanders, across the North Sea, nearly 3,000 people died over thirty weeks in the same year—out of a population of 25,000.

There was more to come. In 1319, a rinderpest epidemic attacked Europe's cattle.[16] This virus, harmless to people but deadly to animals, killed 65 percent of England's cattle, sheep, and goats. People turned to fast-breeding pigs, but rising demand soon led to a shortage. Flocks and herds were so badly nourished that stocks did not recover until 1327. Milk production collapsed to a mere 170 liters per cow. The lack of milk added to the scourge of malnutrition. This was bad enough, but rinderpest also decimated the oxen used to pull plows. Some people had horses, but they were more expensive to feed. Since successful farming depended on putting more land under the plow, the consequences were disastrous.

The Great Famine brought European farmers to their knees. Peasant society was robust enough that it could survive, despite serious food shortages. With their traditional expertise, they could endure several poor harvests. Between 1314 and 1321, six years of rain and famine resulted in political and social turbulence, rebellions, and almost constant violence and warfare. This constellation of misfortunes, at a time when the warm centuries had ended, resulted in a food crisis that was beyond comprehension in the High Middle Ages. The disaster would have consequences for the church, the state, and the entire future of European society, which, when compared to the previous centuries, would take on a tougher and more violent edge—including the horrors of the Hundred Years' War (1337–1453).

Medieval crop yields in Europe were always low, even in the face of what one could call normal weather, which could include unseasonal frosts and autumn hail. This is quite apart from the damage wrought by birds and rodents or, indeed, the pressures of feeding the (previously) booming medieval population. In fact, the population figures speak for themselves. In 1066, William the Conqueror invaded an England where between 2.6 and 3.4 million hectares were under cereal cultivation. These fed about 1.5 million people fairly easily. By the closing decades of the thirteenth century, England's population of 5 million people had to subsist off only 4.6 million hectares, much of it marginal land.

The Medieval Climate Anomaly was not four centuries of warm, settled climate, as many people assume. There were indeed decades of superb weather with weeks of bright sunshine, ample rainfall for the harvest, and milder winters than was often the case. These centuries were an anomaly, remarkable not so much for their warmer temperatures as for their climatic variability, which often swung between extremes of cold and warmth. One certainly cannot claim, as climate change deniers do, that the MCA was warmer than today. Most of the time, average temperatures appear to have been about the same as twenty-first-century norms, with occasional years, even decades, when they were slightly higher. But the effects were nuanced, and the four centuries of the MCA were times of constant climatic variability, as were the centuries both before and after.

Originally, the MCA was considered a European phenomenon. Today, we know that its effects were subtle but global, and at times catastrophic, especially when prolonged droughts settled on semiarid regions like the American Southwest and Southeast Asia. In Europe, the anomaly, with its often bountiful crops, fostered what are commonly known as the High Middle Ages, with their great cathedrals and endemic warfare as people fought for control over resources. Above all, as we shall see in the chapters that follow, the subsequent centuries saw a dramatic increase in vulnerability to climatic shifts as populations rose and more and more farmers were forced to exploit what were, at best, marginal lands for sustainable agriculture. The Medieval Warm Period reminds us once again that sustainability and resilience depend on forward thinking, detailed knowledge of the environment, and long-term planning for short- and long-term climate change.

In spring 1316, generations of increasing vulnerability came to fruition. Spring rain just kept falling, washing seed from the fields, eroding vulnerable hillsides. Malnutrition and starvation-related diseases descended on a wide swathe of Europe for five years. The famine was like the coming of the biblical Third Horseman of the Apocalypse mounted on a black horse, carrying his fateful scales that symbolized the cost and abundance of food. And in the mythic Horseman's footsteps came plague and centuries of cooling as the Little Ice Age ushered in constant, often bitterly cold climatic fluctuations that affected people living in Europe and the Americas.

12

"NEW ANDALUSIA" AND BEYOND

(1513 CE to Today)

It all began with the Norse, long before Christopher Columbus landed in the Bahamas. The first contacts between Europeans and Native Americans came during the Medieval Climate Anomaly when the pack ice retreated in northern seas between Iceland and Labrador in Canada. By 874 CE, Norse colonists had taken advantage of favorable ice conditions in northern seas. They had settled permanently on Iceland, at the threshold of the Arctic. The heyday of their voyaging extended over about three centuries, during warmer and more stable conditions in the eastern North Atlantic (see map in Chapter 8, p. 147).

Eirik the Red, banished from Iceland because of "some killings," colonized Greenland in 986 CE. Soon the settlers had crossed to Baffinland, now part of northern Canada. Leif Eirikson, Eirik's son, sailed as far south as the mouth of the Saint Lawrence River and wintered in northern Newfoundland, perhaps at the L'Anse aux Meadows site at the extreme north of the island, where archeologists have discovered the only known Viking settlement in North America.[1] Numerous voyages to Labrador followed, marked by irregular contacts with Inuit communities and trips to harvest timber, in short supply in Greenland. For generations, Greenlanders paid part of their church tithes to their homeland in walrus ivory from these voyages. In 1075, a merchant named Audun even shipped a live polar bear from

Greenland as a gift to King Ulfsson, something that would have been impossible in the colder centuries after 1200.

The Norse never settled in North America, deterred in part by fierce encounters with Native Americans. But they remained in Greenland for three centuries during generations of colder temperatures and frigid conditions in the western North Atlantic. Alpine glaciers in Baffinland across from Greenland approached their maximum extent between about 1000 and 1250. In addition, temperatures from the Greenland Ice Sheet Project 2 core display cooling episodes between about 1000 and 1075 and again between 1140 and 1220.[2] The Norse population gradually declined until their settlements were deserted by 1450. Exactly why the Norse left Greenland remains a subject of controversy. Increased isolation, a drop in the walrus ivory trade, and perhaps hostile Inuit may have contributed to the abandonment. Only Norse epics preserved memories of an era when Native Americans and Europeans met for the first time.

MYTHICAL "NEW ANDALUSIA" (1513 TO 1606 CE)

The frontiers of Europe's known world expanded significantly during the late fifteenth and early sixteenth centuries. Christopher Columbus and his successors colonized the tropical Caribbean. Aztec Indians were paraded before the Spanish court. Spanish conquistadors explored Florida and New Mexico with disastrous consequences, experiencing intense cold. English colonists founded Jamestown in Virginia, afflicted by drought and low temperatures. John Cabot's voyage to Newfoundland in 1497 and later ventures made it clear that any Northwest Passage to Asia would pass through ice-strewn, very cold landscapes.

Between 1605 and 1607, the Danish monarch Christian IV sent three expeditions in search of the vanished Norse colonies. All ended in failure, suffering through intense cold, even in summer, when ice extended far off the Greenland coast. Thereafter, whaling became the major Dutch preoccupation in arctic waters. The real northern gold lay in Newfoundland's cod fisheries, but an effort to colonize the island by Humphrey Gilbert ended in disaster in 1583.[3] Some of the coldest weather of the Little Ice Age militated against any permanent colonization of Newfoundland. The main focus of attention shifted south of Cape Cod.

As in Europe, the Little Ice Age was never a persistent deep freeze or just centuries of cold weather. The ever-changing climate of these centuries strongly influenced the history of colonial America as well.[4] Freezing winters, prolonged droughts, hurricanes, and great storms were the backdrop that drove ships off course and wrecked them. North America was especially puzzling. European farmers in unfamiliar environments expected the well-defined seasons they were accustomed to, not the dramatic climatic extremes of hot, humid summers and subzero winters. Furthermore, hunting and fishing involved unfamiliar species.

European attitudes of the day to North American weather assumed that climate was constant at any latitude around the world.[5] Classical authors had divided the known world into bands called *climata*, hence the word "climate."[6] *Climata* tended to refer to air temperatures, which vary in a relatively predictable way with latitude. Europe's oceanic climate is humid, with adequate rainfall year-round, relatively small daily and seasonal temperature ranges, and generally minor variations in the onset of each season. This meant that those promoting colonization assumed that those living in North America would enjoy the familiar temperate climates of western Europe. This seemingly commonsense assumption was completely wrong.

Eastern North America, with its extremes of hot summers and cold winters, has a continental climate, governed by air masses that approach over land rather than the Atlantic, which strongly influences European climate. Not only that, but the kind of temperatures enjoyed in Europe lie between forty and sixty degrees north, whereas the equivalent in North America is at thirty-five and fifty degrees north. London is at fifty-one degrees north, at the latitude of northern Newfoundland. Chesapeake Bay in Virginia is at thirty-seven degrees north, the latitude of Seville. Virginia's rainfall occurs mainly in summer and is less dependable, with irregular drought cycles. Climatic reality was a harsh one for European settlers expecting a temperate, warm, Mediterranean paradise, what some authors encouraging potential settlers called a "New Andalusia."[7]

The first reports of lands to the north to reach the Spanish colony in Puerto Rico spoke of an island named Bimini. Spanish explorer

Juan Ponce de León sailed along its coasts in 1513 and renamed the mysterious land Florida. After two unsuccessful expeditions, he gave up ambitious colonization plans, complaining of an unsuitable climate and "very savage and bellicose" people. For the next fifty years, others arrived and returned bitterly disillusioned. La Florida was no "New Andalusia" that would supply the homeland with olive oil and other commodities found in Mediterranean lands. Most rain fell during the summer months, bringing little or no moisture to germinate winter crops. Nor was there a dry season to ripen them. Year after year, crops rotted in Spanish fields. Florida also suffered from savage hurricanes and very cold northerly winter winds. Expeditions large and small traveled as far west as the Mississippi and modern-day Texas, one of them Hernando de Soto's disastrous invasion of 1538 to 1543, remarkable for its suffering and violence. Spain's failure resulted in part from incompetence and poor leadership, as well as unrealistic ambitions. This was no carefully planned imperial exercise with continued financial support. Everything relied on private enterprise that relied on the wealth of the Spanish aristocracy. The royal treasury could not afford such plans.

Colonization also fell apart because of harsh climatic shifts. The southeastern United States cooled significantly during the Little Ice Age, with temperatures as much as 1°C to 4°C lower depending on the location. This cooling resulted in part from cold, dry air and winter snow from the northwest, especially in the sixteenth and seventeenth centuries. The shift in atmospheric circulation with its frigid winds are reflected in Spanish accounts, as are colonists' experiences with severe droughts.[8] Unaccustomed cold, heavy rainfall, and snow left them, whatever their location, in danger of hunger and sickness, as well as attacks by hostile Indians. In 1528, the cold on the Texas coast was so intense that even fish froze in the sea, while it snowed and hailed on the same day. Three decades later, the Hernando de Soto expedition set up camp near the Chickasaw people in modern-day Mississippi in 1541. It was so cold that "the whole night was passed in turning from one side to the other without sleeping, for if they were warmed on the one side, they froze on the other."[9] With all these experiences of severe Little Ice Age cold, virtually unknown today, the dream of "New Andalusia" evaporated. As for the droughts, tree ring sequences, described below, show they were the worst for centuries.

Efforts to settle permanently continued. Spanish captain-general Pedro Menédez de Avilés arrived in Florida in September 1565. He drove out French settlers from Fort Caroline on the Saint Johns River and founded settlements at Santa Elena and St. Augustine just at the moment when a severe drought in the 1560s descended on the colonies, as did a major hurricane. Within six years, half his soldiers had perished of hunger and sickness. The local Indians drove the Spanish out of Santa Elena. Another serious drought during the early 1580s coincided with a brutal war with the local Guale Indians. The Indians surrendered; Santa Elena was rebuilt. La Florida recovered briefly until 1586, when Francis Drake raided St. Augustine, burnt its 250 houses to the ground, and carried off everything. But the city survived, subsidized by the authorities in New Spain, to become capital of Spanish Florida for over two hundred years.

Spanish America now had a legendary reputation for its wealth from the gold and silver exported from Mexico and Peru, which attracted pirates and privateers in significant numbers. Greedy eyes settled on Spanish domains and the virtually unknown coastline north of Florida. The Elizabethan adventurer Walter Raleigh sent two ships on a reconnaissance in May 1584. They landed at Hatteras Sound, then sailed north to Roanoke Island, where the Secotan Indians welcomed them. The visitors returned home with glowing reports of fertile soils, plentiful timber, and even wild grapes. As for the local Indians, they were gentle and certainly not hostile. They were said to live "after the manner of the golden age."

Another expedition to Roanoke, under Richard Grenville (or Greenville) and Ralph Lane, set sail in 1585.[10] Thanks to stormy weather, shipwrecks, and some casual privateering, along with the grounding of the flagship on the Outer Banks and the loss of all their provisions, the colonists arrived at Roanoke in disorder. While Grenville sailed back to England for fresh supplies, Lane remained with about a hundred settlers. The colony soon faltered. Contrary to earlier reports, the soil was thin and far from fertile. Crops died in the fields. Bald cypress tree rings reveal that the most severe drought in eight hundred years persisted at the time of settlement from 1587 to

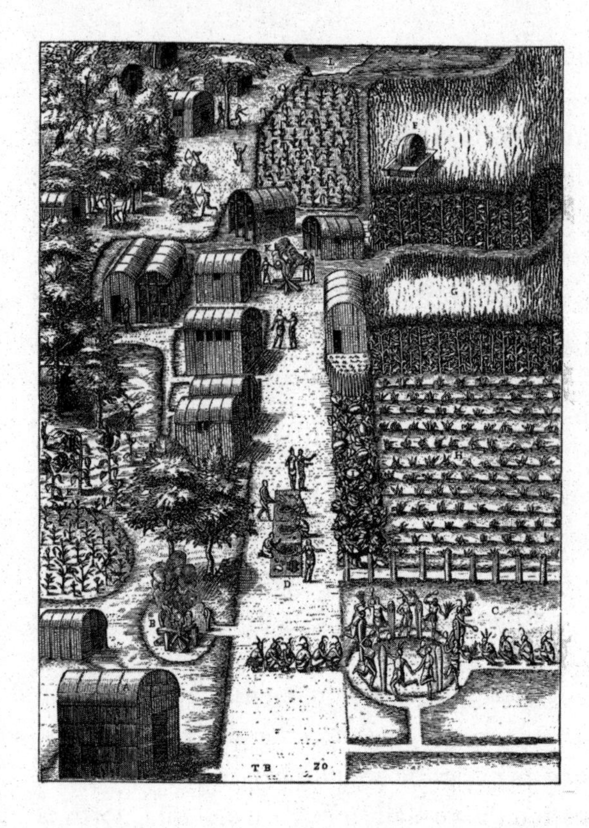

Artist John White (1539–1593) sailed with Richard Grenville to Roanoke in 1585, when he drew this view of Secotan, an Algonquian Indian village. *Alpha Stock/Alamy Stock Photo.*

1589.[11] Also short of food, the Indians refused to trade their corn. No relief ships arrived from England. The desperate colonists, terrified of Indian ambushes, had to forage for provisions. Lane then allied with the local chief's enemies and killed him. Less than a week later, Sir Francis Drake arrived with a plunder-laden fleet but a crew much reduced by disease. He offered to help relocate the colony, but a major gale lasted four days and threatened to cast Drake's ships ashore. The settlers promptly abandoned the outpost and sailed back to England leaving fifteen men at Roanoke, who disappeared without trace. A persistent legend surrounds the vanished settlers, whose fate is still unknown. Most likely, they either joined a local Indian community or were massacred. We will probably never know.

Despite the Roanoke catastrophe, North America and its indigenous inhabitants fascinated people back in England. Enthusiastic boosters planned new colonies, among them the optimistic Richard Hakluyt, a minister and amateur geographer, convinced of the great potential of overseas exploration and trade in English hands.[12] He waxed enthusiastically, and inaccurately, about the abundance of gold and silver, of pearls and rich tropical foods in North America. Rumors of the expanding Spanish Empire in the Americas reached Europe in fits and starts, for the Spaniards considered their discoveries state secrets, accessible to only an elite few. No one in England had read the *Relaciones geográficas de Indias*, compiled in the 1570s and 1580s, which never received widespread distribution. This monumental report included detailed accounts of weather conditions, priceless information for anyone contemplating a voyage of exploration in the Caribbean, northward up Florida, and along more northerly shores. Apart from his geographical errors, Hakluyt reiterated the myth that the entire East Coast from the Carolinas to Maine enjoyed a Mediterranean climate. He wrote of temperate climate, fertile land, warm temperatures, and a place where farmers could raise olives, grapes, citrus, and an abundance of other crops imported to England at vast expense from the Mediterranean. This was a land with "the climate and soile of Italie, Spaine, of the Islands from whence we receive our Wines and Oiles."[13]

A tempting prospect to be sure and one that formed a central part of the Virginia Company's "Instructions by Way of Advice" for the three ships dispatched to the East Coast in 1606. The organizers had learned little from past mistakes elsewhere. They assumed that the climate of their destination would be much the same as that of their homeland, despite the increasing cold of the late sixteenth century.

TROUBLE AT JAMESTOWN (1606 TO 1610 CE)

Three ships and some 144 settlers sailed from London, landing on the East Coast in December 1606. On May 6, 1607, they entered the mouth of the James River in what is now Virginia, where local "Indians" attacked them, but they continued exploration. Eventually they built a triangular fort on a marshy peninsula about eighty kilometers upriver. The low-lying site was strategically sound, with soil "more fertill than can be well exprest." But the fort was close to water's edge, with no freshwater supplies except from the river. Forest pressed on the settlement, so ambushes were a danger. Nothing prepared the colonists for the terrible suffering that lay ahead.[14]

Hakluyt's plans were based on the best information available, but they took the long view. He fixed his eyes on the far horizons of colonization. First, however, came a more pressing problem: surviving the first few winters at Jamestown with only subsistence agriculture to keep people alive. From the beginning, food was in short supply, as the Indians did not provide the generous supplies everyone had assumed would be forthcoming. Soon there was disease and death, with fifty fatalities by August. No one knows what caused the casualties, but famine-related diseases were certainly part of the equation. There was worse to come, as the climate pressed on the colonists. Tree rings provide a sobering account of changing temperatures. By chance, the colonists arrived at Jamestown at the beginning of a long drought, which lasted from 1606 to 1612. The climatologists have also studied sediment cores from the Chesapeake Bay, which show that these were some of the coldest years of the entire millennium, as much as 2°C cooler than those of the twentieth century.[15] Both tree rings and cave deposits in West Virginia reveal a major change in seasonal conditions at the beginning of the seventeenth century, confirming the settlers' own accounts: winters were colder and summers drier.

The extreme climatic shift wreaked havoc on the Jamestown colony during its earliest, most vulnerable years. The hot, dry summer months destroyed growing crops. The James River's water level fell so drastically that its waters turned saline and profoundly unhealthy. No one had thought to dig a well to acquire freshwater. The poor harvests followed by unfamiliar winter cold aggravated both food shortages and personal relationships between the colonists. People subsisted on a pint or less of wheat and barley meal a day, plus any other food they could scavenge. They had both firearms and fishing equipment but apparently made little use of them. Living conditions were austere at best, with many people sleeping on the cold ground. Most likely, as historian Karen Kupperman has argued, the settlers were constantly hungry and in shock, a condition she compared to that of mistreated prisoners of war.[16]

Apart from having to make do with filthy and quite salty water from the river—it took the people two years to dig a well for "sweete water"—they may also have brought typhoid with them from the unsanitary ships. Given the constant threat of Indian attacks, they may also have drawn water a dangerously short distance from where they disposed of their waste. Arguably, many may have perished due to drinking water rather than beer. Most barley harvested in England went into beer, with many people drinking about six pints a day—an important proportion of their calories. The malt from beer was part of daily sustenance. Unprepared for the rough living conditions and poor food, they suffered devastating psychological consequences.

The local Native American villages lived under the Powhatan confederacy, a powerful chiefdom that controlled numerous villages, with about 15,000 people living upstream of Jamestown. Unlike the newcomers, they had centuries of experience of the local climate. Along with all Virginia Indians of the time, the Powhatan combined agriculture with hunting, fishing, and foraging for plant foods.[17] Theirs was a diversified food quest. In spring, they harvested fish from their weirs and trapped small animals like squirrels. Planting season was May and June, while they mostly subsisted on acorns, walnuts, and fish. Others dispersed into small camps and lived off a broad array of foods, both fish and crabs, some hunting, and a wide range of plant foods. June, July, and August were months of relative plenty, as they fed on the roots of tocknough, berries (arum root), fish, and green maize. Late summer and early autumn were a time of harvest and abundance before they preyed on deer and other game through the winter. Some chiefs and more elite individuals managed to store corn for year-round use, but most of the Powhatan planted enough crops to last a few months, then lived off wild foods for the rest of the year.

Judging from the staple carbon and nitrogen isotopes from Native American skeletons of the day, most seventeenth-century Indians relied heavily on maize for their diet.[18] And, despite their intimate knowledge of the resources in their environment, their bones also reveal evidence of episodes of acute malnutrition. Life was never easy, although they had more options than the European immigrants, in part due to their deep understanding of their environment and climate. They could shift their gardens to warmer, south-facing slopes and plant more cold-resistant crops than maize. Under extreme conditions, the locals either moved elsewhere or reverted entirely to hunting and gathering. As anthropologist Helen Rountree has pointed out, the women may have been reluctant to store larger stocks of maize for fear that chiefs and the elite would seize them as tribute.[19] At the time, the Powhatan were living in ever-larger, more concentrated and fortified villages as powerful chiefs competed for power and prestige. From the Indian perspective, the question of what to do about the newcomers was a simple one: How did one maximize the European presence without taking unnecessary risks?

The Powhatan were eager to obtain exotic European metal tools by trading them for maize and other foods. The exchanges ebbed and flowed, thanks to the delicate, often subtle issues of status and diplomacy. By the autumn of 1607, hardly any land had been cleared. The newcomers were living in crude pit dwellings, many of them so depressed that they sat around and did nothing. When two supply ships arrived in January 1608, a fire promptly destroyed almost everything they brought, including the fort. The winter was exceptionally severe, so much so that the James River almost froze across to its other bank. A disastrous relief expedition in 1608 brought more settlers, but their food supplies were minimal.

With some four hundred people crowded into the fort in the face of increasingly hostile locals, there was minimal planting. Inevitably, starvation ensued. About 240 people lived in Jamestown during late 1609. The following summer, only sixty were still alive, the dead buried in a nearby cemetery; their skeletons show clear signs of starvation. By midwinter it was too cold to wade in the shallows in search of oysters. Some desperate Europeans dug up corpses and ate them. The skeleton of a teenage girl found in a cellar inside the fort bore clear signs of butchery; even her skull had been cut open to remove the brain.[20] In 1610, the sturgeon, a critical food, failed to arrive in Chesapeake rivers, perhaps because the persistent drought had made the water too salty. The settlers loaded their ships to leave, only to meet a new, amply provisioned fleet from England at the mouth of the James River, without which the colony would not have survived the Little Ice Age.

NUNALLEQ KNOWS (SEVENTEENTH CENTURY CE ONWARD)

Jamestown's food crisis erupted during some of the coldest years of the Little Ice Age. Even in warmer years, the settlement was vulnerable to crop failure and plagued with uncertain food supplies traded from the Powhatan. The local Native Americans had survived rapid climate change for centuries, yielding to extremes by pursuing a diverse food quest in an environment with fish, wild plant foods, and small game. While some give-and-take was built into the culture, compared with that of the colonists, the locals were far more flexible in the ways they acquired food. And the Powhatan were only one of many Native American tribes who used food diversification to survive the fluctuations of climate of the Little Ice Age.

The Thule village of Nunalleq lies near the village of Quinhagak in Kuskokwim Bay on the Alaskan coast of the Bering Sea.[21] During the fourteenth to nineteenth centuries, the climate there was significantly colder, with greater snowfall, summer temperatures about 1.3°C cooler than today, and much expanded sea ice. Indigenous people lived at Nunalleq for about three hundred years. The densest occupation was during the early and middle seventeenth century—the same time as Jamestown—at the height of the Maunder Minimum, the coldest years of the Little Ice Age. The settlement lay close to rivers that teemed seasonally with migratory fish and with some of the largest concentrations of migratory waterfowl in the world. Small mammals abounded; whales fed close offshore, and sea mammals were plentiful. Meat supplies came from caribou, which fed near the coast during the winter. Today, this is an arctic climate with snow and cool, wet summers. Nunalleq's rich pyramid of food resources also provided raw materials for everything from clothing to hunting weapons. Most important of all, the village's foods were available within a relatively short distance of the site. Food storage, using natural permafrost refrigeration, would not have been a problem. Dietary stress would have been virtually nonexistent. For this reason, permanent occupation at this one location endured for many generations.

The village's strategic location enabled the people to draw on a broad range of foods in a highly flexible way. The variety of foodstuffs at their doorsteps and a high potential for efficient storage meant that one could change the focus of one's hunting when the climate changed simply by focusing on different prey, as climatic shifts are unlikely to have the same effect on all animals in the landscape. Even rapid fluctuations would not have been a problem, especially since foods like salmon were relatively predictable. Risk management would always have lain in the background of the food quest, but it was far easier here than in many environments where the passage of the seasons, droughts, and very low temperatures had a direct effect on food supplies. Currently, dissolving ice, rising sea levels, and higher temperatures melting the local permafrost are washing away the site.

Nunalleq flourished in what one can call a "resource hotspot." Here the locals developed a flexible, varied subsistence strategy bolstered by an intimate knowledge of the local environment. Their sophisticated technology was perfectly adapted to the food quest and life in subzero temperatures, which allowed the village inhabitants to live at one location over the difficult decades, when weather conditions changed abruptly and without warning and food sources changed from year to year. As with the Powhatan, effective and flexible buffering and coping mechanisms allowed the community to survive the worst climatic extremes of the Little Ice Age at a time when competing groups vied for food resources. Theirs was a favored location, which may be why it was ultimately attacked and then abandoned in the late seventeenth century.

DROUGHT BECOMES MEGADROUGHT (LATE 1500S TO 1600 CE)

Finally, we move to the Southwest, where we described earlier how Native American societies adapted to prolonged droughts with mobility and by maintaining kin ties with neighboring communities, near and far. These droughts resulted from natural climatic variability. So did those that descended on the settlers from New Spain, who pushed north into the desert landscapes of New Mexico with its dramatic climatic extremes. They arrived during the late 1500s, one of the driest and coldest periods of the Little Ice Age in the West. For many centuries, Ancestral Pueblo societies had adapted brilliantly to landscapes where crop failures were routine and survival depended on meticulous exploitation of springs and rainfall. Pueblo skeletons reveal frequent episodes of malnutrition, chronic anemia, and short life expectancy in societies where violence was commonplace.[22] The earliest European experiences in New Mexico played out almost identically to those in eastern North America. False hopes, inaccurate predictions, and unfamiliar climates all played their parts, as did the sheer bad luck of severe droughts and other climatic anomalies, caused in part by the Huaynaputina eruption of 1600.

Here, as elsewhere, profound mistrust, incomprehension, and conflict bedeviled relations between Native Americans and newcomers. It's striking just how little knowledge Europeans acquired about local landscapes and foods or about hunting and fishing strategies from their Native American neighbors. They learned from their own hard experience, using their technologies brought from home. How much less they would have suffered had they looked and borrowed techniques to make adequate cold weather gear, waterproof fishing garments, and footwear to combat frostbite developed by the local populations over thousands of years of engagement with the land and its climate.

LOOKING AHEAD

More than almost anywhere else in North America, the Southwest gives us hints about a future being transformed by anthropogenic global warming. While a low level of El Niño activity may be a major factor in megadrought forcing over the Americas, new research combines 1,200-year tree ring reconstructions of summer soil moisture with hydrological modeling and statistical evaluations to show that the nineteen-year period between 2000 and 2018 was the second driest since the year 800. Furthermore, no less than 47 percent of the severity of the current megadrought is the result of anthropogenic climatic warming. Human activities have raised temperatures, lowered relative humidity, and killed millions of western trees. So what would have originally been a routine drought cycle has turned into a megadrought, the second-most severe and sustained for 1,200 years. The symptoms of severe drought are on every side: much reduced snowpack, lower river flows, less groundwater, and increased wildfire activity, among other things.[23] The climatologists attribute the droughts to cool sea surface temperatures in the eastern Pacific—La Niña–like conditions during a time of low El Niño activity. These fostered an atmospheric wave train in the western North Pacific, which blocked storms from reaching the Southwest. The second-worst megadrought of the past millennium began in 2000 and is still in progress. It has now eclipsed the great Dust Bowl aridity of the 1930s and a major drought in the southern Great Plains in the 1950s. We have, of course, no means of predicting if the drought will break with a new cycle of more abundant rainfall soon, but the threat of even greater anthropogenic warming is frightening, for it shows us just how powerful a grip we now have on global climate.

What of the future? At the time of writing (2020), there are no signs of cooling or more plentiful rainfall. Climatic modeling predicts that the drought conditions may worsen by the mid-twenty-first century. The climate change data now available paints a far more complete picture of past droughts caused by atmospheric and ocean anomalies—by natural climatic variability. Those who claim that climate cycles always happen have harped on the twenty-first-century warming as natural. But based on scholarly study of the past in the American Southwest, the 2000–2018 soil drying, enhanced evaporation and early snowpack loss, and the area affected by aridity superimposed on climatic forcing, all amplified by human decisions and activities, have turned what would have been a routine dry spell into a megadrought. And the drying may have hardly begun. Even if natural forces end the current drought, humanity's greenhouse gas emissions across the world will heavily influence the magnitude of future dry spells. We have yet another forcible reminder of the importance of sustainability. Memories are short, but we have seen how groundwater supplies in the past can be drawn down catastrophically in a brief time. This is already happening in countries like India. Construct more reservoirs to store more water? While we have seen this as a short-term solution in some cases, it is delusional to think that this will solve the long-term prediction of less and less rainfall for the future, especially when our actions accelerate that trend.

13

THE ICE RETURNS

(c. 1321 to 1800 CE)

They called it "the Grote Mandreke" or "the Great Drowning of People." Northern Europe was a stormy part of the world in the late thirteenth and much of the fourteenth centuries. No fewer than twelve great tempests roared ashore in the Low Countries sweeping everything before them. Then came the Grote Mandreke on January 16, 1362, which formed as a severe southwesterly gale in the North Atlantic, then swept over Ireland and England, causing the wooden spire of Norwich Cathedral to collapse into the nave below. This was just the beginning. Violent winds and steep-sided waves raced across the North Sea, then crashed into northern Germany and the Netherlands, carrying everything before them. The megastorm demolished over sixty Danish parishes, bowling over cattle like ninepins. A contemporary observer wrote, "The wind tore up anchors and deprived ports of their fleets, drowned a multitude of men, wiped out flocks of sheep and herds of cattle . . . [A]n infinity of people perished."[1] With virtually nonexistent sea defenses and little advanced warning, thousands of people living close to sea level were helpless in the face of what appeared to be divine wrath unleashed to punish sinners.

The Medieval Climate Anomaly petered out rapidly around the time of the Great Famine of 1316 to 1321 with its torrential rainfall and constant climatic swings. Some frigid winters froze large rivers and blocked shipping in the Baltic. Very warm summers were not unknown; nor were severe drought cycles, which could last a decade or just one or two seasons. The great gales that descended without warning were part and parcel of decades of rapid climate change, often with extremes of cold and warmth that ushered in the Little Ice Age.

Climatically speaking, a traveler through Europe during the Little Ice Age would have experienced few differences from today, except for occasional very harsh winters and cycles of intense summer heat. Today, many of us live through bouts of frozen highways or a few weeks of snow or summer temperatures over 20°C. European farmers of the fourteenth century might diversify their crops to protect against frost or dry weather, but they were basically powerless in the face of capricious climatic swings. Acutely aware of the vulnerability, they lived in fear of crop failure and hunger, of diseases caused by malnutrition. The threat of divine vengeance and of the Last Judgement hovered unseen over town and countryside. Then the bubonic plague arrived.

THE BLACK DEATH (1346 TO 1353 CE)

The infamous Black Death descended on Europe between 1346 and 1353.[2] Some twenty-five million victims perished in western Europe—the exact number is uncertain. This frightful pandemic was the second invasion of bubonic plague, the first being the Justinian Plague of 541–542 (see Chapter 5). The villain was a bacterium, *Yersinia pestis*, which infects fleas living on ground rodents, among them the central Asian marmot and various rat species. Exactly when *Y. pestis* first arrived in Europe is unknown, but it was at least as early as 3000 BCE, though the first episode did not cause a true pandemic.[3]

The medieval Black Death originated in central Asia, perhaps in Kyrgyzstan—the land-locked Silk Road country that borders Kazakhstan, China, Tajikistan, and Uzbekistan. From there, the plague spread to China and India. The disease may have traveled along the cosmopolitan Silk Roads or by ship to the Black Sea. By the end of 1346, reports reached European ports that India was depopulated and Tartary, Mesopotamia, Syria, and Armenia covered with dead bodies. Thirty Genoese traders in galleys from Kaffa in the eastern Crimea are said to have introduced the plague to Sicily in 1347. From Italy, the plague spread northwest across Europe. Notable symptoms included buboes (boils under the arm pits or in the groin), fever, and vomiting of blood. Recent DNA analyses from Black Death victims in London and on the Continent have shown conclusively that *Y. pestis* was the culprit.

Why did the Black Death take hold in central Asia? Did climate change play a role in its spread? One way to test this question is by investigating not rats but gerbils. Population densities among gerbils in Kyrgyzstan vary with the prevailing climatic conditions. Warm and wet conditions increase the rising population density of these great gerbils—and their fleas. If the same weather develops over a large region, then plague spreads rapidly. The density of fleas per gerbil increases, plague becomes more prevalent, and, more importantly, the fleas seek out other hosts, including humans and their herds. If temperatures fall and it is much drier, then gerbil populations collapse, and the number of fleas shrinks.

To test this idea, a team of researchers compared tree ring sequences, with their chronicles of rainfall and temperature, from juniper trees in the Karakorum mountains with historical records of plague outbreaks.[4] They found that it took about fifteen years for an outbreak in Asia to arrive in European ports. The rate of spread in more densely populated Europe was much faster than in central Asia: about 1,300 kilometers a year. Epidemiologists and historians have long assumed that the Black Death arrived in a single incident. The new climatological evidence argues that there were climate-driven, intermittent pulses of new strains of the plague arriving from wild rodent reservoirs in Asia as the gerbil populations and subsequent flea populations fluctuated with the climate. There were no such reservoirs in Europe itself.

The consequences were devastating. In Scotland, the sick "dragged out their earthly life for barely two days." Meanwhile, the population fell by two-thirds in and around Paris. One estimate has it that France's population declined by a staggering 42 percent. Many of the dead were exceptionally vulnerable to infection, having suffered from malnutrition during the Great Famine. By the beginning of the fifteenth century, some 3,000 French villages lay abandoned. Poor harvests and wet weather compounded food shortages, aggravated by constant warfare triggered by the Anglo-French Hundred Years' War. Peoples' desperation reached a collective low point between 1420 and 1439, when a high North Atlantic Oscillation (NAO) index brought unusually heavy rainfall. Though there were many fewer mouths to feed, there were still food shortages and famines, many caused by the constant fighting.

The recurrent plagues and irregular famines kept population in check for decades. Many food crises coincided with exceptionally cold winters caused by high pressure over Scandinavia, especially during the 1430s, with their seven years of prolonged frost and severe storms, especially in the Bay of Biscay and North Sea.

Grain production skyrocketed as farmers returned to land abandoned during the Black Death after its end in 1451. Serious recovery also came with the end of the Hundred Years' War in 1453. Temperatures were warmer with ample rainfall. Seven decades later, the 1520s saw five exceptional English harvests until a cold spell in 1527 caused wheat shortages at Christmas and threats of food riots against the rich. Nevertheless, the ancient routine of subsistence farming continued, based on notions of self-sufficiency and crop diversification. But the reprieve was short lived. Bitter climatic winds were gathering on the horizon.

THE LITTLE ICE AGE (C. 1321 TO THE LATE NINETEENTH CENTURY CE)

The so-called Little Ice Age was a "snap" of significant cooling that occurred after the Medieval Climate Anomaly but was not a true, long-lasting ice age. François Matthes, a respected glaciologist, who served on the Committee on Glaciers of the American Geophysical Union, used the term for the first time when writing in 1939: "We are living in an epoch of renewed but moderate glaciation—a 'little ice age.'"[5] Matthes used the term in an informal way—he didn't even put the name in capitals—but today it is a well-established climatological label.

Back in 1939, the Little Ice Age was little more than an idea. Today researchers have accumulated both proxies and historical records for the Little Ice Age from all over the world, not only from Europe and North America but from as far afield as Australia, Oceania, and Japan. For instance, well-documented records of the blooming dates of Japanese cherry blossoms go back six hundred years and provide ample records of cooling. A recent global temperature reconstruction uses no less than seventy-three different worldwide proxies that confirm the cooling, especially from 1500 to 1800. Today the Little Ice Age stands out as the most pronounced climatic anomaly since 6000 BCE, not, of course, including today's anthropogenic global warming.[6]

What actually happened? Between about CE 1250–1300 and 1850–1900, the world became slightly cooler for reasons that still elude us. Deep sea cores around Greenland, Iceland, and Labrador provide a well-documented portrait of an arctic sea ice pulse moving southward as temperatures dropped abruptly. For instance, well-dated, high-resolution ocean cores from the East Iceland Shelf document an abrupt cooling for about sixty to eighty years after 1300, the result of arctic ice moving southward. There was a brief warming in the mid-fourteenth century, followed by renewed abrupt cooling in the late 1300s. After another period of less ice invasion, there was generally more ice from 1500 to the early 1900s. Whether these ice changes resulted from volcanic events or solar change or from other factors is currently unknown.

"Slightly cooler" is very much an average term as the cooling trend varied over time and space. Serious global cooling began around 1400 and only ended in about 1850, when greenhouse gasses caused by industrial pollution cancelled long-term orbital forcing (the effects of slow changes in the tilt of the earth's axis and shape of its orbit around the sun, which can involve the redistribution of solar energy in terms of both latitude and seasons).

The Little Ice Age climate was never static. Shorter forcing episodes, like volcanic eruptions or variations in solar activity, had temporary effects but did cause blips in the climatic record. Other extremes included the catastrophe of the Great Famine, as well as megadroughts, exceptionally severe winters, cycles of major gales, and other episodes that had profound effects on human societies, among them plague epidemics, crop failures, and the periodic decimation of flocks and herds. Such events aggravated short-term vulnerabilities and slowed human resilience.

Eyewitness accounts of the early cooling are rare. Wouter Jacobszoon, a Catholic abbot at Gouda in the Netherlands, moved to Amsterdam in 1572. He kept a diary of endemic violence and Catholic persecution but also complained about the cold. Even staple foods like grain and herring became unaffordable in Amsterdam. The snow lasted until April. As if it were still winter, the cold continued. In November 1574, a gale caused a flood that ruptured dikes and turned flooded fields into ice-covered deserts. In Prussia, a protestant pastor, Daniel Schaller, wondered if the end of the world was nigh. "There is not only a great scarcity of bread, and very dear prices for our beloved corn and grain . . . The wood in the forests no longer grows as it did in the old days . . . [T]herefore *ruina mundi* [the ruin of the world] must be at our gates."[7]

Jacobszoon and his ilk beseeched the Lord for help again and again, but to no avail. Tree ring records from these years do indeed show that tree growth had slowed. Ten earth tremors had shaken Prussia since 1510. The devout Schaller considered them portents of the impending "Last Judgement and the last earthquake in which all dead wake up and come out of their graves before Christ's judgement."

But the Last Judgement never arrived. Instead, climatic shifts continued, and as the oceans cooled, soon enough herring appeared in the North Sea in vast numbers, to the delight of fisherfolk. But the cold persisted. The river Thames had frozen over in London five times between 1408 and 1437 and twelve times between 1565 and 1695. (The last time the Thames froze over was in 1963, during the coldest January since 1814.) These were years when frost fairs thrived on the Thames. Some enterprising vendors even roasted whole oxen on the ice. Winter temperatures not only fell but were more unpredictably extreme. Temperatures reconstructed from proxies confirm many more frozen periods of the Rhône River during the fourteenth century and from the late sixteenth to nineteenth centuries than in earlier ones.

Hendrick Avercamp (1585–1634), *Ice Skating in a Village*, c. 1610. Avercamp was painting for a lively market in winter scenes.
Everett Collection/Alamy Stock Photo.

The Little Ice Age in late-sixteenth-century Europe was not a happy time, with widespread social disorder, often caused by rising grain prices. More than seventy food riots unfolded in Britain alone between 1565, the year after William Shakespeare's birth, and 1660. In earlier centuries, English vintners had exported wine to France, but their harvests evaporated in the face of the cold. Wars, sporadic famines, and bitter cold affected the lives of millions of Europeans. France suffered greatly, both from constant warfare and from bad harvests caused by the cold. At least four million people died due to military violence, starvation, and epidemic diseases during the late sixteenth century. In 1590, King Henri IV, the Protestant monarch, besieged Catholic Paris. He could not acquire sufficient artillery, so he decided to starve the city out. A savage winter had wreaked havoc with food supplies; angry mobs demanded food, but the defenders held on. The dead, together with desperately hungry people, too weak to move, lined the streets. By the time Catholic soldiers broke the siege in August 1590, 45,000 people, a fifth of the city's population, had died of starvation or succumbed to disease.[8] It was no coincidence that emigration from Britain and Europe accelerated during these times.

BALTIC GRAIN AND DUTCH INFRASTRUCTURE (SIXTEENTH CENTURY CE AND LATER)

Change was afoot. The first climatic innovations came in Flanders and the Netherlands as early as the fourteenth and fifteenth centuries.[9] The Baltic states and the Ukraine had long been a granary for much of Europe, the grain exported through Amsterdam and as far south as Italy. Early in the seventeenth century, 75 percent of grain imported from the Baltic states arrived in Amsterdam and was stored in huge warehouses. Domestic grain production had become uneconomic.

In response, both Dutch and Flemish farmers experimented with growing animal forage and cultivated grass for their cattle. They planted peas, beans, and nitrogen-rich clover on land previously left fallow. As more fallow land went under cultivation, animal husbandry assumed ever-greater importance. Manure, meat, wool, and leather entered the marketplace, as the new agriculture broke the slavish dependence on grain and offered a new domestic trade. The farmers planted clover on previously cultivated cereal plots, and their cattle grazed on the meadows before owners replanted with cereals. This self-perpetuating farming cycle made the land far more productive, especially when crops included turnips or hops for brewing beer, along with purely industrial crops like flax and mustard.

Not that Baltic trading was easy: far from it. Ice was a perennial difficulty, especially in frigid winters. On February 12, 1586, in the middle of a severe winter, high winds and freezing temperatures trapped eighteen ships off the port of Hoorn in rapidly spreading ice. The townspeople broke through the ice with axes and laboriously brought the ships into port. Winter storms were even more dangerous. On September 9, 1695, a procession of gales wrecked dozens of ships across the North Sea. About 1,000 mariners perished. The Dutch coast was fully exposed to the prevailing westerly winds of summer. Caught off this hazardous lee shore in high winds, many merchant ships ran aground.

Amsterdam merchants in their comfortable dwellings and warehouses responded quite effectively to the challenges of Little Ice Age winters. But the sailors who carried their cargoes endured serious hardships and often lost their lives. As historian Dagomar Degroot remarks, "Many Dutch citizens adapted to and exploited an environment in flux. They may not have been aware that their climate was changing, yet consciously or unconsciously, they responded in ways that benefitted their interests, and in turn those of their society."[10] All this was despite the constant vicissitudes of war and the ever-complex diplomatic maneuvering that complicated trade relations across the Baltic. For instance, when wheat was in short supply, cheaper rye, which yielded less desirable bread, came into wide use. Prices fluctuated for both rye and wheat as a result. Far from being cowed by these challenges, in times of scarcity, Dutch merchants drew down the huge grain stocks held in Amsterdam, charging high prices for cereals, especially rye, exported to famine-threatened areas to the south plagued by crop failures.

There was more to their entrepreneurial adaptation. Holland was a network of waterways large and small, of canals, rivers, lakes, and coastal passageways, in addition to tracks on land. Diverse and closely connected, Dutch transportation networks made it much easier to move around than elsewhere in Europe, except during the most severe storms of the Grindelwald Fluctuation (1560–1630), with its colder temperatures, and the Maunder Minimum.[11] Amsterdam and Hoorn developed a ferry service of small sailing vessels that departed at set times and arrived at various places, whether empty or full. This "turn ferry" system was so successful that it came into widespread use in the coastal provinces during the sixteenth century. Two centuries later, no fewer than eight hundred turn ferries left Amsterdam weekly for 121 destinations across the Dutch Republic. Head winds and gales could cause chaos, but the system worked remarkably well. In 1595, Fynes Moryson, a wealthy Englishman on the first stage of a long journey to Jerusalem, traveled from Leewarden to Groningen in "boisterous" winds. His party traveled in a private ferry. Propelled by a fearsome but favorable westerly wind, the passengers narrowly escaped shipwreck when they lost their rudder in "a great tempest of wind."

Both city governments and merchants developed new canals with tow paths, used by barges towed by horses. Contrary winds were never a problem, and people could travel between Amsterdam and Hoorn in about two hours at a leisurely speed of seven kilometers an hour. More than 300,000 passengers used "pull ferries" by the mid-seventeenth century, traveling in first- or second-class compartments. Children traveled at half price.

People, including slaves, plus basic commodities, even hay and fish, and also letters traveled in small boats owned by both farmers and entrepreneurs. These *schuiten*, some under sail, were up to ten meters long and traversed not only major waterways but all kinds of minor canals and channels that reached small communities. When the weather was warmer, the ferries usually ran smoothly. But ice and sustained freezing could stop ferry traffic for up to three months in severe winters, playing havoc with the transport of dairy products such as milk, which were a major ferry commodity. Even then, local ingenuity kept shipments and people moving, giving the republic a dramatic advantage over countries like Britain and France, where infrastructure away from fresh and salt water was a much greater challenge.

The Dutch domestic transportation networks gave travelers a flexibility and resilience that allowed people to move in the face of rapid climatic shifts of the Little Ice Age. Gales and ice were major hazards for Baltic and North Sea trade. Fortunately, the variety in the Dutch diet meant that food shortages were virtually unknown, despite constant shifts in grain prices.

The diversified agricultural economy made it easier to adapt to short-term, sudden climatic shifts, especially with easy access to Baltic grain, and the inland waterways facilitated the transport of food to almost anywhere. These infrastructure improvements occurred alongside massive land reclamation that caused farmland in the Netherlands to expand by about 100,000 hectares between the sixteenth and early nineteenth centuries, most of it reclaimed between 1600 and 1650. Fortunately, the Dutch enjoyed a flexible social organization that encouraged smallholdings at a time when farmers' incomes were rising. At the same time, younger families sought more than the basics. Housing improved dramatically as brick construction became widespread and consumer goods like clothes and furniture more readily available.

Competent and highly competitive, Dutch and Flemish farmers were unique in a Europe still dominated by subsistence farming that had changed little since earlier centuries. The innovations spread gradually. By 1600, English market gardens near London grew vegetables for the city market. Sixty years later, Dutch immigrants introduced the cold-resistant turnip to eastern England with its lighter soils. Green turnip tops were a useful substitute for hay. The low-lying eastern English fens had long been a refuge for herders, fisherfolk, and fowlers. Over the seventeenth century, Dutch-born engineer Cornelius Vermuyden, an expert on sea defenses, reclaimed over 155,000 hectares of fenland and turned it into some of the most productive arable land in Britain.[12]

Experimenting with new crops became another strategy to diversify subsistence. Maize and potatoes, imported from the Americas, became commonplace. A Spaniard returning home from South America in about 1570 brought the latter to Europe. At first, potatoes were merely a botanical curiosity, even considered an aphrodisiac, one anonymous authority remarking that eating a potato "incites to Venus." The exotic tubers were far more productive than oats and other crops and were rich in mineral content. They became animal fodder before becoming a staple in Ireland and across Europe during the eighteenth and nineteenth centuries. New crops, innovative farming methods including extensive manuring, and improved drainage, as well as a policy of enclosure, slowly broke England free of the tyranny of cereal farming. France was not to throw off that yoke for two more centuries. Meanwhile, addictively compelling products like tobacco and chocolate became part of the social hierarchy.

Meat consumption also soared. By the eighteenth century, the English had developed an insatiable appetite for beef, mutton, and pork. In 1750 alone, London butchers slaughtered at least 74,000 fattened cattle and 570,000 sheep. As the productivity of crops improved and fodder became more plentiful, herds became larger, valued for their flesh, hides, and by-products. Stock husbandry became an art during the eighteenth century, notably in the hands of Robert Bakewell, a farmer from central England, who bred cart horses for load hauling and cattle for the quality of their meat. His greatest success was with sheep, especially the "new Leicester," a breed that matured so rapidly he could bring them to market in two years.[13]

SUNSPOTS, VOLCANOES, AND SIN (1450 CE AND LATER)

While farmers and preachers still invoked fearsome images of the hoary nightmare that linked climatic disasters with divine wrath, the seventeenth and early eighteenth centuries also witnessed major scientific advances, many of them in astronomy. Astronomers recorded transits of Venus and Mercury and established the velocity of light by observing the orbits of Jupiter's satellites. Some of their studies helped our understanding of how the cosmos affects climate on earth. They studied eclipses of the sun and published the first detailed studies of the sun itself, in addition to an examination of sunspots.

In 1711, English natural scientist William Derham remarked on the low level of sunspot activity between 1660 and 1684. This led him to remark, "Spots can hardly escape the sight of so many Observers of the Sun, as were then perpetually peeping upon him with their Telescopes."[14] Everyone assumed until 1774 that sunspots were clouds that covered the sun, so there were few new observations until the nineteenth century. Today, we know that sunspots are places where the sun's magnetic field rises from below its surface. Sunspot activity waxes and wanes about every eleven years but does not affect us directly. Days, even weeks, can pass without any sunspot activity. Over the past two centuries, only the year 1810 saw no sunspot activity whatsoever. By any standards, the low level of sunspot activity during the Little Ice Age was unusual. Whether or not these lulls caused lower temperatures during that period is unknown, but they coincide in considerable part with its coldest years.

There were three minima. The first longer cold phase of the Little Ice Age came between 1450 and 1530. This coincided with a low level of sunspot activity known as the Spörer Minimum, named for a German astronomer.[15] The Spörer years were cold, but those of the second minimum, dubbed the Grindelwald Fluctuation for a town in the Alps, lasted from the early 1560s to 1620 and were significantly colder. During the coldest Grindelwald years, the growing season in northern Europe became as much as six weeks shorter. Many farmers switched from wheat to more cold-resistant barley, oats, and rye. Nevertheless, crop failures still occurred, especially on marginal lands. The Maunder Minimum (1645–1715) was a period of very low sunspot activity that coincided with a period of lower-than-average temperatures in Europe and North America. The Thames in London and Dutch canals froze over. During the Maunder, the sun emitted weaker ultraviolet light, leading to less ozone in the stratosphere. The decrease caused planetary waves, which kicked the North Atlantic Oscillation into a negative mode. Winter storms tended to be colder under such conditions, and temperatures were lower, as confirmed by limited historical records.

Sunspot activity did not cause the Little Ice Age. Most likely volcanic activity was a major player, for the cold increased with intensified volcanic activity. Mount Huaynaputina in southern Peru burst apart on February 19, 1600, the largest eruption of the past 2,500 years, dwarfing both Pompeii and even Mount Tambora and Krakatoa during the nineteenth century (see Chapter 14).[16] The Huaynaputina explosion threw thirty cubic kilometers of ash and rock thirty-five kilometers into the atmosphere. Ash fell like rain over hundreds of square kilometers. Volcanic ash mantled the volcano-surrounded town of Ariquipa. A local scholar, Felipe Guáman Poma de Ayala, remarked that one could see neither sun nor moon nor stars for a month. The summer of 1601 was the coldest since 1400 throughout the Northern Hemisphere. Summer sunlight in Iceland was so dim that there were no shadows. Both the sun and moon were little more than "faint, reddish" apparitions. At least four other volcanic episodes during the seventeenth century produced significant cold spikes, but none of them as severe as that caused by Huaynaputina.

Chamonix, now a fashionable ski resort, was then a poverty-stricken village where ice constantly threatened growing crops. The community lost a third of its land between 1628 and 1630 in the face of an assault of avalanches, floods, and advancing ice. Only one harvest in three reached maturity in fields that were under snow for most of the year. In desperation, the villagers persuaded their community leaders to brief the bishop of Geneva on their plight. They told him of the ice threats and of their fear that they were being punished for their sins. The bishop led a procession of three hundred people to four villages under siege from four glaciers. He prayed over and blessed the ice sheets. Fortunately, the blessings seemingly worked, and the ice slowly retreated. Unfortunately, the newly exposed land was too barren for agriculture. Not that the retreat was permanent. Whenever the ice advanced once more, renewed devotions from Chamonix and elsewhere ascended to heaven. Alpine glaciers were much larger than today until about 1850, when their retreat began.

Meanwhile, food prices rose in the face of constant poor harvests and ever-more expensive wines. Crop failures, famines, and resulting diseases led to bread riots and social disorder. As had been the case for centuries, priests proclaimed the persistent bad weather a consequence of God's wrath over a sinful humankind. A frenzy of hysterical accusations erupted during the cold years in 1587 and 1588. Neighbor accused neighbor of witchcraft. The authorities in the German town of Wiesensteig burned no fewer than sixty-three women accused of witchcraft at the stake in 1563.[17] Witchcraft only faded into the background when scientists began to develop natural explanations for climatic events. Until then, God and the forces of the supernatural were ready villains.

ACROSS THE OCEANS (SEVENTEENTH CENTURY CE ONWARD)

For all the revolutionary changes on farm and homestead in response to challenging climatic conditions, some of the most radical changes were in long-distance sea trade. Whereas the Portuguese and Spanish had historically led the way, the Dutch now stepped into the breach, surprisingly, during a period of increased storminess.[18] Severe storms were four times more common in Flanders during the Grindelwald Fluctuation than previously. Most significantly, wind directions and velocities changed substantially with interesting consequences.

The tightening grip of colder conditions during the Grindelwald Fluctuation interfered with ambitious attempts by Dutch mariners and merchants to open an arctic passage across the north of Europe. Ice conditions were formidable and the route too expensive to be viable for long-distance trade. Instead, they turned their attention to small companies that invested in a southern route via the Cape of Good Hope to Asia. For them, the passage to Southeast Asia was hazardous and lengthy, the risks unacceptable. Accordingly, the States-General of the Netherlands brought the companies together to form the Dutch East India Company (Vereenogde Oostindische Comagnie [VOC]) in 1602. The company, virtually a conglomerate, rapidly prospered, trading spices and textiles from India and Southeast Asia for precious metals. The Heren XVII ("Seventeen Lords") ran the VOC, whose ultimate aim was to undermine competing Spanish commerce. In 1619, the VOC's governor-general in Asia, Jan Pieterszoon Coen, seized Batavia (now Jakarta) in Southeast Asia, which became the center of Dutch enterprise in the region. VOC became an enormous enterprise, with more than 30,000 employees, in addition to the labor of a vast number of people from Africa whom it exploited as slaves. The Dutch soon dominated trade between Europe and Asia and between Asian ports for generations.

VOC depended on long ocean passages by fleets of East Indiamen on a scale that minimized risk. Much depended on the company's experience of conditions at sea, especially prevailing currents and trade winds. These habitually blow from the northeast in the Northern Hemisphere and from the southeast in the Southern. Initially, the company's captains experimented with different routes, but the Seventeen Lords established standardized routings through the English Channel, then south to the Cape of Good Hope, and from there east to the Australian coast and ultimately north to Southeast Asia. Two fleets departed annually, a Christmas Fleet in winter and an Easter Fleet in spring. The return passage from Batavia left between November and January, arriving in the republic by November the following year.

By any standards, VOC voyages were dangerous enterprises, especially during the coldest decades of the Little Ice Age, with their frequent gales. Any ship loss was a catastrophe, crowded as they were with both people and valuable merchandise. More than half the shipwrecks due to weather occurred in the North Sea during the very cold decades.

A treasure trove of VOC ships' logs has produced a new understanding of the influence of climatic changes on voyages from year to year. A low NAO and Siberian High (persistent high pressure to the east) increased the prevalence of easterly winds in the northeastern Atlantic during the Maunder Minimum, usually one of the slowest parts of the voyage. The Intertropical Convergence Zone had also moved south, making it uneconomic to stop at intermediate ports. At the same time, the rising strength of the trade winds, driven by upwelling in the southern Caribbean after 1640, increased the speed of VOC ships across the Atlantic. The Little Ice Age shortened the outward journey and increased profits, while the summer monsoon, although weaker, allowed ships to trade through Southeast Asia if they arrived in time.

Dutch merchants and their mariners may have coped better with the cold decades of the Little Ice Age, for the coastal provinces of their nation benefitted greatly from their trade on the world's oceans. The VOC's influence faded during the late seventeenth century, when smaller ships owned by Scandinavian, French, and English owners became faster, carrying other commodities such as coffee and tea, which were more lucrative and elite commodities. Climatically, the decline of the Maunder Minimum increased westerly winds over the northeastern Atlantic, which slowed down outbound ships.

The Dutch Republic had a unique form of political structure mostly governed by councils of urban merchants. Their number included ambitious entrepreneurs, innovators, and—as the Dutch and, indeed, most of the colonial West now acknowledge—extreme exploiters of indigenous Africans. They used their subsequent wealth to improve the technology of land reclamation, shipbuilding, and even firefighting. Rapidly growing Amsterdam became the commercial and financial heart of Europe, a cosmopolitan center for export and import, famous for its commercial efficiency. Above all, the Dutch adapted successfully to the challenges of exceptional cold, making the most of unique opportunities of all kinds.

Ultimately, the Dutch, whether engineers, farmers, mariners, or farm laborers, not only grew accustomed to constant climatic vicissitudes but devised ingenious ways to plot courses through decades of often savage cold and unpredictable natural challenges—all facilitated by the untold efforts of human slave labor. One could call this entrepreneurial capitalism that helped solve environmental challenges. But, in the end, as we'll see in Chapter 14, a gigantic volcanic eruption in 1815 exchanged the equation for everyone.

These events unfolded during centuries when Christian doctrine maintained a religious stranglehold on thinking about nature, the environment, and human origins. Abrahamic religious dogma proclaimed that the story of God's creation of the earth and humanity in Genesis was historical truth. Archbishop Ussher of Armagh, Primate of All Ireland, used biblical genealogies to calculate that God created the earth and humanity on October 22, 4004 BCE. A formidable scholar, Ussher published his finding during a period of decades of major scientific advances in everything from astronomy, biology, mathematics, and medicine to plant classification—to mention only a few. Science thrived in the field, the laboratory, and the study. Agricultural diversification and selective animal breeding took hold; rational argument and discourse competed with religious ideology.

During the Little Ice Age, the long-held assumptions that climate change resulted from God's wrath over human sin faded at a time when rational discourse and careful observation fostered scientific inquiry of all kinds. This was a momentous turning point in the study of ancient and contemporary climate, when science gradually moved to center stage in the study of forecasting climatic conditions. A minority of conspiracy theorists and religious devotees aside, the debate that pits science against other explanations is long over. Paleoclimatology is very much a twentieth- and twenty-first-century science that has transformed our knowledge of global climate. But the primacy of science as opposed to secular and religious speculation began to develop during the coldest period of the Little Ice Age—with fundamental significance for today's and tomorrow's world.

14

MONSTROUS ERUPTIONS

(1808 to 1988 CE)

Colombian astronomer Francisco José de Caldas was mystified. Starting on December 11, 1808, he observed a persistent, stratospheric "transparent cloud that obstructs the sun's brilliance." His observations went further: "The natural fiery colour [of the sun] has changed to that of silver, so much so that many have mistaken it for the moon."[1] A physician in Lima, Peru, noted unusual afterglows during sunsets. These two eyewitness accounts are the only firsthand records of a major eruption, probably in Southeast Asia, that affected temperatures over a broad area of the world. The only other record is a sulphate spike in Antarctic ice cores five years before the great eruption of Mount Tambora, also in Southeast Asia, in 1815.

The mysterious eruption was not alone. Between 1808 and 1835, there were at least five major tropical volcanic eruptions, during several decades when April–September temperatures cooled by about 0.65°C compared with the proceeding thirty years of warmer temperatures.[2] This significant temperature drop was likely related to the intense volcanic activity. Alpine glaciers expanded. Temperature changes generated by these volcanoes reduced monsoon activity in India, Australia, and Africa, bringing drought reflected in low Nile floods and East African lake levels. The Atlantic-European cyclone track moved southward after the eruptions, a move related to less intense African monsoons.

Volcanic activity was one reason why the last phase of the Little Ice Age was remarkable for large climatic fluctuations that lasted for one or more decades. The rapid temperature increases that occurred after the volcanic activity subsided reflect the recovery of the global climatic system from a rare series of eruptions, also perhaps from some limited anthropogenic warming related to the beginnings of the Industrial Revolution. But from the late eighteenth and early nineteenth centuries onward, human-caused increases in greenhouse gasses dominated long-term climatic trends as the Little Ice Age gave way to prolonged warming.

The years of volcanic disruption were also a time of social and political tumult. Volcanoes, with their primordial lava flows and cataclysmic explosions, became fashionable spectacles. The explosive crater of Italy's Vesuvius was a popular tourist destination and a highlight of the grand tour. Less affluent pleasure seekers witnessed spectacular volcanic paroxysms in London pleasure gardens and theaters. "The Eruption of Vesuvius Vomiting Forth Torrents of Fire" promised one newspaper advertisement in what became a competitive business.

FRANKENSTEIN'S ERUPTION (1815 CE)

Vesuvius was, and still is, small volcanic business compared to the massive eruptions that emanated from the Pacific "Circle of Fire" in Southeast Asia. Ice cores from the Arctic and Antarctic have revealed a major eruption in the southwestern Pacific region during 1808 (the date is still being finalized), which was the third-largest eruption since the early 1400s, behind only Tambora (described below) and the eruption of Mount Kuwae on Vanuatu in the southwestern Pacific in 1458. The 1808 eruption produced cooling as far away as England, with snow lingering on the lowland hills of Scotland throughout the spring. In Manchester to the south, morning May temperatures were below freezing. The weather was relentlessly cloudy for weeks during the summer of 1810.

A large volcanic eruption influences global temperatures for one or two years, which is very different from the effects of a series of them, including the 1808 event. The world had already cooled from the 1808 eruption before Mount Tambora on Sumbawa Island in

Southeast Asia blew into space in 1815, the most violent volcanic event in modern times. Tambora had long been dormant, but today we know that it had exploded before, 77,000 years ago, impacting lands far beyond Asia. Like its remote predecessors, the 1815 cataclysm was a truly global event.

After weeks of rumbling, Tambora erupted on the evening of April 5, 1815. For three hours, huge flames and clouds of ash burst from the mountain. Five days later, the volcano exploded, glowing brightly as lava flowed down its slopes. As many as 10,000 people perished in a morass of flames, ash, and molten lava. Two or three days later, Tambora collapsed into itself, a six-kilometer-wide caldera replacing the former summit. The mountain lost a kilometer and half of its height in the explosions, which could be heard hundreds of kilometers away. Ash mantled ships to a depth of a meter or more. Cinder-filled clouds turned daylight into darkness. A tsunami caused by the explosion wreaked havoc ashore and resulted in numerous casualties. Islands of pumice stone from the eruption floated as far east as the middle of the Indian Ocean. Over a radius of six hundred kilometers, darkness descended for two days. The landscape was unrecognizable; fields were ruined. Thousands starved as the impact of the disaster intensified. Sumbawa Island was deforested and has never fully recovered. Folk memories of the eruption remain vivid. The locals call the April 1815 Tambora eruption the "time of the ash rain," and with good reason.[3] Internationally, the environmental and social consequences of the Tambora eruption extended far into the future. The ash discharge was one hundred times that of Mount St. Helens in Washington State in 1980. The 1883 eruption of Krakatoa, also in Southeast Asia, the first major eruption to be studied systematically, reduced direct sunlight over the earth by 15 to 20 percent.

Immediately after the eruption, ash drifted at will in the stratosphere. The enormous volcanic cloud, with its sulphate gases, formed aerosols that became dense enough to reflect solar energy back into space, creating a hotter stratosphere but cooling surface temperatures. The thermal synchrony between land, ocean, and the heavens was disrupted; the monsoon, with its three-month rains, weakened. Instead of torrential monsoon rains, 1816 brought drought to an enormous swath of South Asia. Temperatures fluctuated wildly; drinking water stagnated in storage tanks; crops could not be planted lest they wither. Moisture stress seriously inhibited tree growth. When the atmosphere recovered in September 1816, the monsoon descended with unseasonal fury, causing widespread flooding.

Half a world away, Tambora lay behind Europe's dismal weather of 1816. The winter was one of cold and violent storminess, and things didn't get any better as the year progressed. In fact, 1816 was dubbed the "Year Without a Summer," an apt label, but one that masks the magnitude not of an isolated event but of a global climatic anomaly.

The English poet Percy Bysshe Shelley and his second wife, Mary, accompanied by the poet Lord Byron, vacationed in Switzerland that unusual summer and climbed in the Alps in "a violent storm of wind and rain." Both they and the locals complained of the cold and almost continual rain, as well as fierce wind and thunderstorms, that confined them to their house. It was the coldest winter in Geneva since records began in 1753, with 130 days of rain between April and September; it even snowed in July. The weather-bound Mary wrote her iconic horror story about a young scientist named Frankenstein, now a literary immortal.[4] Byron composed a poem titled "Darkness" on a day so cold that birds roosted at noon. In that monstrous year, even animal fodder became unaffordable; horses died or were eaten. Over the border in Baden, this situation inspired German inventor Karl Freiherr von Drais to create his "running machine," later called the velocipede, as a substitute for the horse. But his foot-propelled machine, forerunner of the bicycle, endangered pedestrians, so it was banned, even in congested Kolkata in India.[5]

The abnormally low temperatures throughout the growing season devastated not just animal fodder but all harvests. English wheat yields were the lowest between 1816 and 1857, at a time when food consumed two-thirds of a family's budget.[6] The French harvest was half that of normal, partly because of widespread flooding and thunderstorms with hail. The wine harvest began on October 19, the latest in many years. Grain prices rose, but fortunately large grain reserves from previous harvests kept prices reasonably low for a while. Thanks to somewhat improved transportation and grain imports, there was a food dearth rather than a universal famine. Nonetheless, a full-blown food crisis tormented Germany, while Zurich's streets swarmed with beggars. Social unrest, food riots, and violence exploded across a Europe still recovering from the havoc of the Napoleonic Wars.

Stagnation in manufacturing and trade, widespread unemployment, and the stress of a rapidly industrializing British economy engendered widespread rioting put down by militia. Ireland, newly reliant on the frost- and damp-intolerant South American staple, the potato, suffered from widespread hunger in the face of inadequate relief efforts.[7] The subsistence crisis triggered massive emigration throughout Europe, with thousands of hungry, poor people traveling down the Rhine to Holland seeking passage to America. More than 20,000 impoverished Rhinelanders, not to mention English and Irish, immigrated to North America to escape the miseries of subsistence farming on highly fragmented landholdings, where the risks of crop failure were ever higher.

PANDEMONIUM (1815 TO 1832 CE)

Stormy weather continued into the following year. By 1817, the watery environment of the Bay of Bengal had stimulated genetic mutations in the cholera bacterium that lurked in its drought-struck waters. The unusual droughts and floods created by the Tambora eruption triggered a global cholera pandemic that decimated Indians and Europeans alike. (An estimated 125,000 people died in Java alone, more than perished in the eruption itself.) Cholera spread with inexorable disregard for national boundaries. It reached Persia in 1822, Moscow in 1829, Paris in 1830, London a year later, and North America in 1832. The long-term impacts on history were enormous. Cholera exposed the newly interconnected world to the dangers of pandemics and the social inequalities in the crowded, poverty-stricken slums where the disease flourished.[8] The climatic consequences of the Tambora eruption laid the foundations for a pandemic that rivaled the Black Death in its destruction.

The skies over China displayed brilliant colors in the summer of 1816 after the eruption. One observer described "sheets of glowing pink, which . . . darted upwards from the horizon." As environmental expert Gillen D'Arcy Wood aptly puts it, "This much can be said of Tambora's plume: It was an attractive killer. A tragedy of Nations masquerading as a spectacular sunset."[9] The effects were immediate, with record low temperatures in eastern China and substantial crop failures. In Shaanxi in the Northwest, crop failures were so severe that thousands of people abandoned the province, the same reaction as played out in Europe. But the greatest suffering fell on Yunnan in the Southeast, a mountainous region with close ties to trading networks in Southeast Asia. With its fertile valleys wedged between mountains, Yunnan had long been a granary for rice and wheat cultivation. With a temperate, pleasant climate, the province escaped much of the fury of the Indian and Asian monsoons. The intensification of Yunnan agriculture in the late eighteenth and early nineteenth centuries tripled the local population from three million in 1750 to twenty million in 1820.

Yunnan had no spring or summer in 1815, as cold Tambora weather descended just a month after the eruption. The cloudy and rainy conditions decimated winter crops; frost froze rice paddies in August, destroying the rice harvest. A terrible famine afflicted the land from 1815 to 1818 as cold north winds reduced crop yields by two-thirds, or perhaps even more. Temperatures were about 3°C below the normal average. This may seem like a small difference but remember that every 1°C decrease in temperature reduced the growing season by as much as three weeks. Unfortunately, Yunnan's grain reserves were already depleted by a drought in 1814, so there was widespread hunger. Snow fell in 1816, with another rice failure caused by cold temperatures and icy fogs never experienced before. The famine did not ease until 1818, when atmospheric conditions returned to normal.

By early 1817, the central government was alarmed enough by the emergency that its officials dispensed free grain from official granaries. This was nothing new, for Chinese officials had carefully monitored grain prices and distribution for centuries. They brought in grain during the harvest months, then distributed it in winter and spring as local supplies dwindled and prices rose. According to local officials, there was enough grain stored in Yunnan to feed every grown man in the province for a month. But years of granary neglect meant that the system fell part rapidly and people were hungry. They turned to cash crops. There was an explosion in poppy cultivation in Yunnan, which generated a profitable trade in opium. A century later, Yunnan imported almost all its grain from Southeast Asia. By then, China exported over 80 percent of the world's narcotics. The traffic developed in the eighteenth and nineteenth centuries in which Western countries, mostly Great Britain, exported opium grown in India and sold it to China, where it was also grown. The British then used the profits from the sale of opium to purchase such Chinese luxury goods as porcelain, silk, and tea, which were in great demand in the West.

AMERICAN DEGENERATION? (1816 TO 1820 CE)

In the Western Hemisphere, the "Year Without a Summer" has entered historical folklore and was for generations the most written about climatic event in North American history. Many contemporaries called it "Eighteenth-Hundred-and-Froze-to Death." Dust clouds appeared in the skies above Washington, DC, in early May 1816. An intense high-pressure system over eastern Greenland in early May funneled arctic air southward, just as it would do in midwinter. Temperatures plummeted as cold air flowed into New England when a massive trough of low pressure stalled over the Great Lakes. A black frost in mid-May decimated newly planted crops, as did a cold wave that dumped a third of a meter of snow across the Northeast. Freezing temperatures gripped the East as far south as Richmond, Virginia, and west to Cincinnati, Ohio. There were further frosts in June, late July, and August, the only year in which this has ever been recorded. The growing season shrank to a mere seventy days in New Haven, Connecticut; hay was in very short supply for hungry cattle.[10]

The combination of dry weather and unusual cold extended into 1817, when retired president Thomas Jefferson reported that most of his crops were failing. Three years later, he faced ruin, as the crop failures plunged him ever further into debt. Jefferson had always thought of the United States as an agrarian nation, but now the dream seemed threatened. None other than the celebrated French scientist Comte de Buffon, criticized by clergy for his scant references to God's role in climate and nature, declared that the relentless cold of North America could not support crops or any animals other than diminutive species. This was the old argument that latitude dictated climate, so much so that it was said that European settlers "degenerated" in a land that he described as "a perfect desert."

Buffon's theories were, of course, nonsense, but they retained great popularity among wider audiences. Even Mary Shelley referred to the "degenerated" Americas where Frankenstein's monster desired to escape civilization. Weather became a touchy subject for Americans visiting Europe. Jefferson was a vigorous advocate of his homeland during his sojourn as American ambassador in Paris during the early 1780s. His landmark book *Notes on the State of Virginia* launched a frontal attack on Buffon's hypotheses. He defended both the people and the animals, citing the enormous size of the extinct mammoth and Native Americans whose "vivacity and activity of mind is equal to ours." As for the American West, it was a portrait of health and happiness.[11] Jefferson had a passionate, imperial vision of the United States. He met with Buffon for dinner. They agreed to disagree in a thoroughly civilized manner.

As in the seventeenth century, the climatic optimism that pervaded much writing about the United States during the early nineteenth century did not endure in the face of record cold, triggered first by the eruption of 1808, which reduced New Haven's temperatures well below average. Then came Tambora, which mainly affected the Eastern Seaboard, with bumper harvests in areas like Ohio to the west. But the bitter cold of the Tambora event passed through the economy as a depression that lasted from 1819 to 1822. Many people who moved west to flee the depression comprised the first climate-driven mass migration in the Americas and ended up at the mercy of land speculators. Add to these migrants thousands of immigrants fleeing difficult conditions in Europe, and there was an inevitable land bubble and credit crisis. As European crop yields improved dramatically after 1820, the prices of American cotton and wheat tumbled. By then a financial panic had caused over three hundred banks to collapse overnight. In the final analysis, the Tambora eruption did not just cause the breakdown of European markets for American commodities; it undermined the banking system and every sector of the American economy, causing perhaps the most devastating economic crisis of the nineteenth century at a time when only ten million people lived in the United States.

COAL KILLS COLD (1850 CE AND LATER)

When did the Little Ice Age end? The conventional date has long been about 1850, connected to the persistent warming from intensifying industrial activity. Yet, judging from ice cores in the Swiss Alps, it was not as straightforward as that.

At their maximum in the middle of the nineteenth century, some 4,000 large and small Alpine glaciers extended about twice the distance they do today. Then they began to retreat in about 1865. Scientists long assumed that increased temperatures and decreasing rainfall caused their rapid retreat, signaling the end of the Little Ice Age. The assumption proved wrong, for local temperatures were cooler than they were in the late eighteenth and early nineteenth centuries. Rainfall was apparently unchanged. Some other forcing mechanism was at work to cause the mysterious glacial retreat.

High-altitude ice cores drilled about 4,000 meters above sea level revealed dramatic increases in black carbon emissions and carbonaceous aerosols, caused in part by incomplete burning of fossil fuels and other human activities.[12] These passed into the atmosphere with the expansion of the Industrial Revolution, starting in Britain during the mid-eighteenth century, then in France, Germany, and most of western Europe over the next ten decades. Black carbon emissions rose precipitously after 1850. When glacial researchers converted the energetic effects of the soot on glaciers of the day, they found that the melting effect of the black carbon caused the glaciers to retreat rather than engendering any dramatic change in temperatures. Owing to areas of intense industrialization surrounding the Alps, the levels of black carbon climbed rapidly between 1850 and 1870 and increased steadily until well into the twentieth century.

Coal combustion for heating and industrial uses was a significant cause of pollution, as was increased tourist traffic in the Alpine region. The air in Alpine valleys was so sooty that nineteenth-century housewives never dried their laundry in the open air.

More is known about Alpine glaciers than those anywhere else in the world, so it would be a mistake to assume that the end of the Little Ice Age in the Alps coincides with glacial retreats elsewhere. By no means did all glaciers give ground in the 1860s. The Bolivian

Andes experienced glacial retreat as early as 1740, Himalayan glaciers in the mid-1800s, and those in Argentina and Norway in the early 1900s. Like so much else related to climate, changes in temperature and other shifts were as local as they were global.

Nor was the warming in Europe clear-cut after 1850. Warmer years continued through the 1870s, except for occasional very cold Februarys and wet summers after 1875. A cold snap in 1878–1879 brought conditions that rivaled those of the 1690s. Farmers in eastern England were still harvesting after Christmas at a time when cheap American wheat from the prairies was inundating the British grain market. An agricultural depression ensued. This was also a time of sustained monsoon failures in India and China, when between fourteen and eighteen million people perished from famines caused by cold, drought, and monsoon failures. As late as the 1880s, hundreds of London's poor died of accidental hyperthermia in the persistent cold snap. In 1894–1895, large ice floes formed on the Thames in midwinter. Then prolonged warming began. Between 1895 and 1940, Europe enjoyed nearly a half century of relatively mild winters. Only 1916–1917 and 1928–1929 were unusually cold, but with nothing like the sustained chill of the Little Ice Age.

The depressed economic conditions of the 1880s led to a flood of immigration to new lands. Thousands of unemployed farm laborers moved from the countryside to cities or relocated to Australia, New Zealand, and other countries where they sensed opportunity. Nineteenth-century immigration to Australia, North America, New Zealand, South Africa, and elsewhere brought land-hungry European farmers in search of fertile, uncleared land. They descended like locusts, felling millions of trees for cultivation, firewood, and construction lumber for growing towns and cities.[13] The extensive logging added new levels of carbon dioxide to the atmosphere and enhanced warming. A standing forest can retain as much as 30,000 metric tons of carbon per square kilometer in its trees and more in its undergrowth. When the trees are felled, they no longer absorb carbon, so much of it enters the atmosphere. One estimate calculates that the twenty-year global burst of agriculture and land modification between 1850 and 1870 raised the carbon dioxide level in the atmosphere by some 10 percent, even after one takes into account the carbon absorbed by the oceans. Isotopic levels in ancient California bristlecone pines rose during these years, at a time when fossil fuel burning was an insignificant factor in the environmental picture. Compare this with the catastrophic impact of the 76,000 fires set by farmers and loggers in Brazil's Amazon rainforest in 2020. In July 2020 alone, the forest shrank by 1,345 square kilometers.

Coal burning is a major contributor to black carbon accumulation. As long ago as August 14, 1912, a newspaper on New Zealand's North Island, the *Rodney and Otamatea Times, Waitemata and Kaipara Gazette*, noted, "The furnaces of the world are now burning about 2,000,000,000 tons of coal a year. When this is burned, united with oxygen, it adds about 7,000,000 tons of carbon dioxide to the atmosphere yearly . . . The effect will be considerable in a few centuries."[14] Not that this obscure paper was the first to talk about the perils of warming. On July 17, 1912, Australia's *Braidwood Dispatch* had published the identical story a month earlier, copied from Britain's *Popular Mechanics*, which had run a similar story in March of the same year. The dire warnings were nothing new. They had been around in some form or another for a long time.

BURNING ISSUES (LATE NINETEENTH CENTURY CE)

As early as the seventeenth century, Londoners complained of the polluting smoke generated by burning sea coal (bituminous coal found at or below sea level). The perceptive John Evelyn complained of "fuliginous steame" from coal fires. King Charles II considered ways of reducing the growing smog problem but to no avail. In 1843, there were at least five hundred industrial chimneys in Manchester, covering the city with a "dense cloud" through which the sun shone "like a disc without rays."[15] By the 1850s, London was the wealthiest and most powerful city on earth and subsequently the most crowded and polluted. By 1900, 6.5 million people lived in London, a city heated by coal. At the same time, the city had appalling sanitation problems that had turned the Thames into a horrendous sewer. The city's "pea soup" fogs, commemorated by Sir Arthur Conan Doyle in his Sherlock

Holmes stories, were famous throughout Europe and persisted into the mid-twentieth century. Industrial activity combined with natural conditions to create a toxic atmosphere.

An impression of growing air pollution also comes from an esoteric field of research: landscapes in nineteenth-century paintings.[16] J. M. W. Turner (1775–1851) was a landscape artist whose expressionistic studies of light and atmosphere were vivid and sublime. He was among those who painted stunning sunsets during the three years after the Tambora eruption. As Turner once remarked, he painted landscapes to show what a scene was like. Redder sunsets may reflect volcanic events. During the 1970s, meteorologist Hans Neuberger analyzed paintings from 1400 to 1967 in art museums in Europe and the United States. His statistical analysis revealed a slow increase in cloudiness over the centuries, but, after 1850, skies are less blue, the atmosphere hazier, which, apart from artistic convention, Neuberger attributed to increased air pollution that diminished the blueness of European skies. A team at the National Observatory in Athens is now examining the sunsets painted by numerous old masters. However, as environmental historians have pointed out, we must take many factors into account before we can use such works as a reliable barometer of climatic conditions, including the fashions of the art marketplace. Nevertheless, many lesser-known day-to-day paintings of late-nineteenth-century shipping on the Thames feature haze in polluted London skies.

Coal burning and industrial pollution lie behind continual warming, but precisely when human activities began contributing to today's long-term warming is hard to establish. To some extent, it's a matter of definition. For example, the Intergovernmental Panel on Climate Change, founded in 1988, uses the arbitrary date of 1750 CE as when industrial activity began to spread more widely, leading to an increasing use of fossil fuels and greenhouse gas emissions. A more nuanced estimate comes from a synthesis of marine paleoclimatic data, which shows that the coolest ocean surface temperatures of the past 2,000 years occurred between 1400 and 1800, resulting in considerable part from enhanced volcanic activity over the past millennium. In many areas, this long-term sea surface temperature cooling trend reversed during the industrial era, matching equivalent temperature trends on land. Both show warming developing after 1800.

These average-temperature readings mask significant regional temperature differences. Sustained warming began in tropical seas in the 1830s, mirrored by warming on land in the Northern Hemisphere. Warming began some fifty years later in the Southern Hemisphere, especially over Australasia and South America. At issue here is the date at which the effects of climate change fell outside the range of the climatic variability to which natural systems adapt. The latest assessments suggest that the widespread climatic warming that marked the twentieth century and continues today has its roots in a sustained trend that began in the tropical oceans and over parts of the Northern Hemisphere as early as the 1830s. Did volcanic activity play a role? The volcanic cooling caused by Tambora did not endure, instead giving way to an interval of accelerated global warming as the climate recovered. Most likely, the greenhouse forcing of industrial-era warming began by the mid-nineteenth century and continues to this day.

ANTHROPOGENIC WARMING (1900 TO 1988 CE)

The years between 1900 and 1939 witnessed a high incidence of westerly winds and mild winters, characteristic of a high North Atlantic Oscillation phase. The pressure gradient between the Azores and the Icelandic low was steep enough to maintain the prevailing winds. Air temperatures over the world reached a peak in the early 1940s, with notable warming in regions close to the Arctic like Iceland and Spitsbergen. Pack ice in the north shrank by about 10 percent; mountain snow levels rose; ships could now reach Spitsbergen for five months a year as opposed to three during the 1920s. Rainfall increased over northern and western Europe, making the western front in World War I a wilderness of mud. The plentiful rainfall lasted into the 1920s and 1930s as warming continued. After 1925, Alpine glaciers vanished from valley floors as they retreated into the mountains. The Oklahoma Dust Bowl of the 1930s resulted from stronger Pacific westerlies that increased the arid wind shadow of the Rocky Mountains far to the east. Changes in atmospheric circulation resulted in more reliable Indian monsoons, with only two partial failures between 1925 and 1960.

During the 1940s, scientists began to talk about the persistent warming, which transcended the normal climatic swings of earlier times. They speculated about retreating arctic ice and the disappearance of the northern pack. But they did not take human actions like deforestation or the use of fossil fuels into consideration, therefore excluding and absolving most human-caused changes. Climatological research was in its infancy, at a time when there were no computer models, satellites, or global weather tracking. Apart from a lack of tools, the constant variations in rainfall and temperature tended to obscure the all-important long-term trends. Carefully organized meteorological data spanning millennia and centuries was also in short supply.

The North Atlantic Oscillation switched to a low phase during the 1960s, as the westerlies weakened and western Europe became colder and generally drier during the winter. Ice completely covered the Baltic in 1965–1966. In 1968, during an exceptionally cold winter, arctic sea ice surrounded Iceland for the first time since 1888. That year, eastern Europe and Turkey experienced their coldest winter in two centuries. Record low temperatures over the Midwest and eastern United States convinced many people that another Ice Age was imminent.

In 1971–1972, the North Atlantic Oscillation flipped. Warming resumed and appeared to accelerate. The Baltic was completely ice-free in 1973–1974. England experienced its warmest summer since 1834. Record heat waves scorched much of western Europe in 1975–1976. More weather extremes, increased hurricane activity, and numerous droughts painted a picture of a very different global climate than earlier in the century. A moment of political truth came in 1988, when a two-month heat wave seared the Midwest and eastern United States. Long stretches of the Mississippi River practically dried up. Barges lay aground for weeks. About half the crops on the Great Plains failed, and over ten million hectares of the American West burned in wildfires ignited in drought-plagued country.

A single Senate hearing in Washington, DC, on June 23, 1988, turned climate change and global warming from an obscure scientific concern into a public policy issue. This was when climatologist James Hanson testified before the Senate Energy and Natural Resources Committee on a day when the temperature reached 38°C, or 100°F.[17] Hanson drew on data from 2,000 weather stations across the world to demonstrate not only global warming over the past century but the dramatic resumption of higher temperatures after the early 1970s. He flatly declared that the earth was warming permanently because of humankind's promiscuous use of fossil fuels. Our climatic future would include a much higher frequency of heat waves, droughts, and other extreme climatic events. His testimony thrust anthropogenic global warming into the public arena overnight. Nothing that has happened climatically since then has shown that Hanson was wrong.

But awareness of climate change slowly passed into the background of public consciousness. Industrial development had both transformed the American economy and led to the development of a sophisticated financial system that did much to insulate most Americans from the harsh realities of crop failures and sudden climatic shifts. Since the 1990s, climate change has moved to the forefront of public consciousness, thanks in considerable part to the ravages of major El Niños and persistently rising temperatures and long drought cycles. That human activities are leading to inexorable global warming is now scientific certainty. It is only now, in a humanly caused warming world, that climate change is rapidly becoming a major issue in global politics, despite the rantings of anachronistic ideologues.

15

BACK TO THE FUTURE

(Today and Tomorrow)

America, Rome, China, India; floods, volcanoes, droughts, balmy years; famine, war, exploitation, adaptation, cooperation. In this book we have told many stories of how our ancestors both successfully and unsuccessfully dealt with climate change. But does the past matter in the context of current climate change? After all, as most agree—other than a few climate change deniers—today's climate is altering because of our own industrial-era behavior, whereas, prior to the 1800s, such change was naturally driven. As a group of climatologists recently emphasized, much ancient climate change unfolded at a local and regional level, whereas today anthropogenic warming and climate change are continuous and global—we can now share information about climate change both globally and almost instantaneously.[1] These immediate connections give each human new power. No matter who you are, you can have an influence on future climatic shifts, in what is sometimes called the "Greta Thunberg effect." So why should anyone look at the ways in which diverse, often unconnected, preindustrial societies adapted to climate change? What possible relevance does the experience of our dead forebears have for the climatic changes we are facing today and will confront even more directly in the future? As novelist L. P. Hartley wrote in 1954, "The past is a foreign country: they do things differently there."[2]

While the world has changed immensely over the 30,000 years described in this book, as have our economies, we and the people who lived during these millennia—regardless of our skin color or our country of origin—have shallow evolutionary roots. We *Homo sapiens* are fundamentally alike: we all have the same hormones, bodies, blood, and cerebral potential. And because we are all the same species, our reactions to unexpected events are often remarkably similar across time and space. We know this because the eyewitness accounts that describe how Romans reacted to the Vesuvius catastrophe sound eerily akin to how people responded to the Mount Tambora eruption of 1815 or the Mount St. Helens explosion in the American Pacific Northwest in 1980. The same types of behavior were likewise displayed during Hurricane Katrina, which flooded New Orleans in August 2005 and Superstorm Sandy, which attacked Cuba and the eastern United States in 2012.

From these natural disasters, we've learned that the most powerful weapons for resilience and survival are qualities that go back deep into the past: the importance of local cooperation in the adaptation and recovery process among communities, kin groups, and quite often even wider groups that might normally be in political, spiritual, or cultural opposition. By looking to the past, we also get to see the full gamut of our species' potential behavior; some of it is horrific and exploitative, but we can learn from these actions too.

New scientific research is also revolutionizing our perspectives on both global and local climate change of the past. A half century ago, we knew precious little about the European and American climates of the past 2,000 years. Now we can decipher seasonal climatic shifts for two millennia or more. Research in China and India, in Australia and New Zealand, and on Pacific Islands is telling us that climate change was a powerful and often inconspicuous driver throughout the human past. We also know a great deal about the present, about the ecological harm we humans have inflicted and continue to inflict on the world's ecosystems. Many climate change researchers predict a hazardous future in a world shaped in large part by booming human populations living at ever-closer quarters and by climate change generated almost entirely by human activities. Correctly, they call for a search for solutions, for reducing human-caused warming. It remains a global problem, not a problem to be muddied by petty nationalisms and partisan politics.[3]

We repeat this call. We must remember that we are one species, with shallow evolutionary roots, representing close connections across the globe, and that we are all players in what went before—and in what lies ahead.

BEING HUMAN

We mention our shallow roots since our modern industrial world is built on the backs of recent slavery and colonialism. To justify the use of slaves and the exploitation of other lands, Western colonialists emphasized the gulf between people from different parts of the world (or "races," an illusory term that defies clear categorization and is largely based on superficial and readily changeable looks). This brainwashing goes deep. Until the 1990s, many human evolutionists argued that modern *Homo sapiens* from different continents had extremely distant evolutionary connections (around the two-million-year mark) and that the different "races" spontaneously evolved in different regions—in China, Europe, Africa, and so on. We now know that our species emerged onto the scene around 300,000 years ago in Africa, becoming fully anatomically (physically and presumably mentally) modern after 150,000 years ago, and that all those who live outside Africa only left that continent around 50,000 years ago.

Yes, there was some subsequent interbreeding with Neanderthals and others, but the inherited DNA is not restricted to a single skin color, hair type, or head shape and most certainly does not create the vast differences espoused by racists. As a species, we are biologically alike. Our outward appearances are skin deep, and looks are readily changed within a generation. But what is universal, what makes us "anatomically modern humans," is our internal wiring: we all have large brains, with the capacity for speech, forward planning, and creative thinking. Such capabilities help define our distinctive identity as *Homo sapiens*. And the key behavioral trait that sets modern humans apart from the rest of the animal world is culture. Culture is a uniquely human attribute and has been our primary means of adapting to our ever-changing environments. But culture has a deep antiquity that long predates our own species.

Much to the horror of staunch Victorians, we are naked apes. Our entire evolutionary ancestry goes back six million years or more, to a time when early human forms split with the ancestors of the modern chimp. Evidence for human culture only appears in the archeological record several million years later, at the 3.3-million-year-old site of Lomekwi 3 in Kenya, in the form of crudely chipped stone tools. This assemblage demonstrates how an archaic human species had already begun to manipulate natural stone into its service. Yes, some other intelligent animals also use tools—the octopus and the chimpanzee come to mind—but not to the degree that we do and have done.

Only humans rely on a wide variety of "material culture" (things we make) rather than on our bodies alone as a buffer between ourselves and the environment. This is uniquely different from other animals, who depend on their fur, sharp teeth, webs, venom, horns, and so forth. Culture is fascinatingly diverse: today's Inuit in the Far North fabricate thick layers of clothing, build igloos, and hunt for food using stone, antler, and horn implements, while most Londoners live in brick homes, wear factory-produced fabric clothes, buy food from supermarkets, and use computers. But don't let this diversity blind us to our inherent similarities.

Despite their variety, all human cultures share one common trait: they adapt constantly in response to all manner of variables. In hunting societies, a caribou herd may alter its spring migration route without warning; relatives in neighboring bands (or streets) may quarrel; information gleaned from other women may cause a group to move twenty kilometers away to exploit groves of ripening fruit. Subsistence farmers may argue over land inheritance and depend on kin living at some distance to supply food in hungry years. City leaders may squabble over trade routes or even start wars to control commodities like iron ore, rice, or oil. All societies have their volatile moments when decisions are made or debated.

Strikingly, humans have oftentimes adapted in the same general ways. This is one reason why mobility has been a powerful catalyst in adaptation. It has been a logical adaptive strategy for thousands of years. But when we look into the deeper past, in the absence of written records, it can be hard to decipher economic, environmental, political, or social changes. The process of adaptation is complex. Archeologists have become pretty adept at tracking major economic and technological changes, like the shift from hunting and gathering to agriculture and herding. While much human behavior lies in the realm of the intangible—we cannot excavate a lost language or long-vanished oral traditions—we can look at the various technological innovations that have helped us adapt to major climate shifts.

Layered clothing tailored with eyed needles protected people living on the Eurasian steppes during the bitter cold of the late Ice Age, though this is not to suggest these Ice Age folk were the first to use the needle; they were not—people in the Sibudu Cave of South Africa were using the needle by 61,000 years ago, for example. Clay pots came into use by around 15,000 years ago to cook and store food. But again, clay—surviving in the form of decorative figurines—had been in use in even earlier times. Humans innovate, but the intelligent ones also learn lessons from the past and from other humans. Bronze, used for axe heads and swords, revolutionized both agriculture and warfare, followed by the even stronger iron, with metallurgical methods soon adopted by communities far and wide. Irrigation technology and city sanitation, chariots and outrigger canoes were all remarkable inventions by our "wise" species. Sometimes these innovations arose independently in widely separated areas (as with crop domestication in, say, the Americas and in the Near East); sometimes remarkable inventions faded and disappeared (as with the end of the Indus Civilization, when comparable sanitation technology would not be seen for another 2,000 years). But sometimes clever ideas were shared across wide areas, from community to community, a positive contagion if you will. We twenty-first-century industrial-era humans did not spring alertly onto the stage of history armed with supercomputers and atomic energy. At least three million years of technological experiment and innovations lie behind us, as do millions of years of human adaptation to climate change.

Why do these legacies persist? Because we have always transmitted our knowledge and experience to the young. Until post–Ice Age warming began, almost all societies thrived in small hunting and gathering bands, where experience was all-important. The cumulative experience of the elders passed orally from one generation to the next, as it also did among all preindustrial farming and herding communities, sometimes by word of mouth or in song, chants, and storytelling, and, of course, by example. Much centered on an intimate knowledge of the local environment and its animals and plants, which provided not only food but also medicine, clothing, and the raw materials for hunting weapons, digging sticks, and other tools. This environmental knowledge resulted from generations of close observation of natural phenomena, which changed with the passing seasons, as well as of game and impending weather, be it a snowstorm, a hurricane, or telltale signs of a dry offshore wind that would decimate growing crops. The know-how was remarkably encyclopedic and speaks to us through the things we left behind: the details of reindeer coats in Ice Age cave paintings, the feathers used to craft Hawaiian chiefly cloaks, or the wild grasses consumed by cattle at different seasons. The Inuit had, and still have, numerous words to describe different snow and ice conditions. So did Aleut Indians paddling their canoes through waves in the Bering Strait. They described these waves with a comprehensive vocabulary. All of this was inherited knowledge, passed from father to son, from mother to daughter, from one generation to the next—and from the ancestors to their descendants.

KNOWLEDGE PASSED DOWN

A remarkable amount of environmental knowledge has come down to us through the generations, little of it committed to paper or parchment and much of it in oral traditions, of which more and more vanish every year. Such understandings of the natural environment acquired over thousands of years should have been our most enduring legacy from the past. Alas, with the advent of industrialization during the eighteenth and nineteenth centuries, this vast repository of critical expertise is rapidly depleting, swept aside by industrialized food production with its fertilizers and by promiscuous forest clearance, marked by an almost total disregard for indigenous peoples and the future of our world.

Despite this, a deep reservoir of traditional climatic and environmental knowledge does still survive, not only in the memories of subsistence farmers but also in long-forgotten anthropological and historical archives. Nineteenth- and twentieth-century Western anthropologists collected much of this information, thanks to their abiding interest (often as the handmaidens of colonialism) in the minutiae of everyday life, which included subsistence agriculture and routine traditional practices. Much of this traditional knowledge revolved around what we now call risk management. Talk to a farmer living from one harvest to the next and working the land by hand or read stories of Victorian fisherfolk venturing offshore under sail in the North Atlantic, and you find yourself mingling with cautious people. Their concern is, and was, long-term survival in worlds where the specter of hunger and malnutrition always lingers on the horizon.

Such people—and there are still millions of them around the world—live in rural communities, not huge cities. A huge gulf of understanding and, into the bargain, a sense of urgency and a need for action separate us urban folk from those traditionally linked to the environment. Our lives are disconnected. Those who live close to the environment have a profound emotional investment in their farmland—witness the difficulties of resettling people forced to move when confronted with major hydroelectric schemes. Subsistence farmers have an intimate knowledge of their land, of their landscape, which is unimaginable to most of us living in crowded urban settings. Their rapidly vanishing knowledge of *local* ecology, of *local* climate changes such as drought cycles and their telltale signposts in the environment—a priceless archive of how small communities survive climate change—is forgotten or ignored. Such knowledge remains a closely held secret among such people as Amazonians, Andean farmers, Pueblo Indians in the American Southwest, and rural communities in central Africa. One can hardly blame them, given the rapacious colonization of recent centuries. Environmental wisdom, much of it far more fine-grained than that painstakingly acquired by today's ecologists, is a compelling, often forgotten legacy from the past. Is it safe to assume that, as the climate crisis intensifies, this gulf will eventually be bridged? In today's rapidly changing climatic world, will we currently disconnected city dwellers one day begin to confront the environment more directly? If we do, the benefits will be enormous.

KIN TIES

Another priceless inheritance from the past: ties of kin (by which we mean actual or imagined family connections within and between communities). No human society has ever been completely self-sufficient, even if many Ice Age hunting groups perhaps encountered a mere thirty people or so outside their bands during their short lifetimes. Even the smallest bands maintained at least sporadic contact with neighbors near and further afield. Sometimes they would meet to obtain wives or husbands, resolve disputes, or exchange hides, exotic ornaments, and other commodities such as toolmaking stone. These contacts depended on ties of kin. Kinship has been a preoccupation of anthropologists for generations, and with good reason, for families, extended families, and links with members of kin groups living at greater distances were always an essential cement for human societies large and small.

Being a member of a kin group carries obligations—commitments of marriage, of support, and especially of reciprocity—the assumption being that fellow kin support one another with food and other commodities when the need arises. Such ties of cooperation and reciprocity were a vital part of handling the risks of crop failure and prolonged drought. Kin ties played a central role in Ancestral Pueblo society, where, for instance, the people of Chaco Canyon maintained strong reciprocal ties with fellow kin in communities at considerable distances, ties that even permitted them to move to these villages if conditions in the canyon became untenable—as they did.

Powerful kin ties were a major element in much more complex preindustrial civilizations. The earliest Mesopotamian cities were ultimately agglomerations of villages, with neighborhoods grouped by kin group membership and occupation. Most Egyptians maintained close ties with rural villages that extended back generations. So did Indus city dwellers in South Asia and Khmer villagers in Southeast Asia. The ancient Maya and the Inca of the Andes relied heavily on kin ties in environments where mountain runoff and rainfall were uncertain.

The legacy of kin is desperately diluted in the vast urban societies of today, where anonymity and isolation are commonplace. There are, of course, numerous exceptions, but it's safe to say that the strongest roots of kinship lie in societies that still maintain close ties to the land. Fortunately, today's most closely knit urban societies include city neighborhoods with a strong sense of cultural identity, as well as such organizations as fraternal societies and churches, which maintain strong links to their local memberships. Some of the most powerful weapons against catastrophic climatic events, such as Hurricane Katrina and other disasters, are, strikingly, ties of kin and community, institutions with long traditions reaching far into the past. Such coping mechanisms are bound to assume even greater importance in tomorrow's reality of more extreme weather events.

MOVING TIMES

Dispersal and mobility are also powerful heirlooms from earlier humanity. In recent decades, we've begun to confront the reality of reduced mobility in a world of much higher population densities and city dwellers in the millions. How, for example, would one leave a city like Houston, Miami, or the central core of Shanghai at short notice and in large numbers? You're talking about an almost impossible task. Indeed, modern nation-states stop people from moving without the correct documentation.

Yet the ability to move freely is our natural state and was the norm of human existence for millions of years. After all, hunting and foraging depended on constant movement—pursuing game, tracking edible plant foods, and following up on vital knowledge acquired from distant lands. When human populations were small and bands no more than a few families, movement was effortless, a highly adaptive way to skirt damage caused by exceptionally heavy floods or by short or long drought cycles. The hunting bands who fished and foraged in the heart of Doggerland (now the North Sea) were constantly on the move in a low-lying environment that changed the landscape rapidly within an individual's short lifetime. The numbers of people involved were small; movement was etched into their daily existence.

The equation changed fundamentally when farmers settled in permanent villages, anchored to their lands, which passed down through generations, thanks to rules of inheritance firmly embedded in kin groups and lineages. In many cases, slash-and-burn farmers moved their villages when nearby gardens were exhausted. Moves occurred perhaps every generation or so, with the settlements often moving in rough ovals that eventually returned them to sites abandoned years before. Most communities were small, so movement was relatively straightforward, a matter of communal decision and public opinion. Under circumstances like these, dispersal was always an option in the face of prolonged drought and other factors like catastrophic storms.

Mobility was an important strategy of resilience, especially in preindustrial societies like the Indus Civilization, where cities maintained strong links to rural communities. Food or water shortages triggered movement to the countryside in search of these resources, often using kin relationships based on long-established reciprocal obligations. The mass migrations of today dwarf even nineteenth-century movements in response to strong El Niños, often propelled by poverty and long-term drought. Responding to such often involuntary population movements raises complex social issues. But dispersal and mobility as tactics to escape climatic shifts go back deep into the past and are near-instinctive human behaviors when confronted by stress. Forced movements were rarer but occurred when conquered peoples were relocated to new and often remote territories, a practice followed by the Assyrians and the Inca, to mention only two examples. Developing global policies to handle ecological refugees is an urgent problem for a world in which mobility is no longer an asset but a liability.

LEADERSHIP

The legacies of human behavior inherited from earlier societies assumed even greater importance in preindustrial civilizations that developed around the world after 3100 BCE. This was the point when leadership assumed a central role in many human societies, which impacted, both positively and negatively, how people dealt with climate change.

Leadership began with experience and acquired wisdom, qualities associated with respected elders and with shamans and spirit mediums, thought to be powerful intermediaries with the supernatural world. The ancestors played a fundamental role in human existence and, once they died, became embedded in supernatural forces that determined the continuity of human existence. It became increasingly powerful in farming societies with close ties to land formerly cultivated by their ancestors. To legitimize ownership and to stake claims to the land, the ancestors were evoked (and still are by every single nation-state).

Kin ties and ancestors: these were the underpinnings of leadership as human societies became more complex. With the emergence of the first preindustrial civilizations, cultural and social responses to climate change assumed new, much more complex dimensions. As villages became towns and cities, hierarchies developed among lineage and other kin groups, with some acquiring perceived spiritual and political authority. Chiefs had long acquired and maintained their clout through personal charisma, as well as their adept use of gifts and appointments to prestigious offices as ways of developing loyal followers. Such loyalty was a fleeting thing, for much depended on gift giving and reciprocity—favors and gifts, be it food, political support, or even military assistance, bestowed in the expectation that they would be returned in the form of loyalty. One kept one's followers contented, or they evaporated and followed another leader.

With hereditary leadership came social inequality, haves and have-nots. Most preindustrial civilizations were centralized pyramids of social inequality, ruled by powerful individuals and their privileged kin at or near the pinnacle. Below them were serried ranks of officials and priests, who supervised and taxed thousands of commoners, whose endless labors accumulated the food surpluses that supported the panoply of the realm. Ancient societies run for the benefit of the few depended on large food surpluses, compelling political and religious ideologies, and decisive leadership for survival. All of them, to varying degrees, were vulnerable to both local and more global climate change. In many respects, they were not too different from many societies today, with their social chasms between haves and have-nots.

In almost every ancient society, subsistence farmers lived close to the edge, with hunger a constant reality and careful risk management an unspoken one. But what happens when a privileged elite depends on reliable food surpluses and the rations extracted from them to survive? The specter of vulnerability raises an even uglier head in the face of unpredictable climatic events—hurricanes and great gales blowing ashore in North America and the North Sea, hundred-year rains devastating irrigation works in Peruvian river valleys, and, above all, drought. Without question, long-term dry periods were the greatest hazard faced by preindustrial civilizations. We've made a distinction between short-term droughts, maybe lasting one to three years, and much more serious hydrological arid cycles that can last a century or more. Village farmers were accustomed to brief droughts, perhaps a couple of bad years that caused people to grow less popular crops or to forage, often in vain, for wild plant foods. There would have been hunger, maybe deaths, but life continued. Such short dry cycles were not catastrophic for early civilizations, especially if the rulers had taken steps to store grain for lean years.

Hydrological drought cycles, or megadroughts, as we now call them, were a different matter. The 4.2 ka Event, the megadrought of 2200 to 1900 BCE, rippled across the eastern Mediterranean and South Asia. Monsoons weakened, Nile floods failed catastrophically in 2118 BCE, and the Egyptian state dissolved into competing provinces. Food surpluses evaporated; confidence in pharaonic authority dissolved. Several generations passed before the state reunited under warrior rulers. There was no longer talk of divine rulers controlling the inundation. The pharaohs now proclaimed themselves "shepherds of the people" and invested heavily in irrigation schemes and state-owned grain storage. Ancient Egypt survived until Roman times.

ORGANIZING RESOURCES

Egypt was fortunate, for its domains, with their fertile soils, lay within secure frontiers, making armed invasions virtually impossible. The state became more resilient and was self-sustaining for the long term, despite endemic factionalism in the pharaoh's court. In an era of short life expectancy and rudimentary medicine, the question of the succession was pervasive as rivals discreetly maneuvered for power. Constant intrigue and slippery alliances were part of every preindustrial civilization, most of which rose and fell with bewildering frequency. The reasons are easy to discern. Nearly every city-state in Mesopotamia, every Maya kingdom, and every early Chinese demesne—to mention only a few—had infrastructure problems. The pharaohs moved armies and commodities of all kinds by water with ready efficiency. In the desert, they relied on donkeys and, much later, camels, but the logistics of feeding their pack animals limited the size of the loads a caravan could transport.

Land-based states confronted a harsh reality, which extended into recent times. Rulers and merchants could only move goods either on human backs or with pack animals like donkeys or llamas. Heavy loads like timber or sacks of grain could be moved on rivers or lakes or even close inshore on the ocean. But in infrastructure terms, commodities of all kinds traveling by land could only be transported about fifty kilometers before pack animals had to be rested or changed. This reality also imposed powerful limits on the size of territory that lay under tight control of the court—perhaps under one hundred kilometers across. Beyond this frontier, control was more nominal and depended heavily on the loyalty of nobles and provincial officials.

The effectiveness of adaptations to sudden climate changes, especially hydrological droughts and weak monsoons, relied on decisive leadership and kin ties. Strong leadership created loyalty and infrastructure, especially in outlying areas, to manipulate ample food surpluses to tide the people over during food shortages. These were critical when crops failed and commoners were hungry. So were kin ties that linked people living in cities like Mohenjo-Daro, Tikal, or Ur to communities in the hinterland. Such links were like an insurance policy, for the obligations of reciprocity allowed hungry people to quietly disperse to more desirable areas and situations when caught by drought—when the Tigris River failed to overflow or months of rain decimated medieval crops in Europe. An ancient survival strategy could pay dividends.

Preindustrial civilizations, including the Roman Empire, depended heavily on human labor, pack animals, and what was effectively monoculture over wide areas of the empire. In its later centuries, the empire depended heavily on imported grain from Egypt and North Africa, transported by large grain ships that shuttled back and forth across the Mediterranean. The infrastructure for feeding much of the empire was about as efficient as was possible when oar- and sail-powered cargo ships and pack animals moved harvests from remote farmlands. The empire depended heavily on slaves for its maritime shipping. In the end, the imperial economy was undermined not by its infrastructure but by weak monsoons that reduced the Nile inundation catastrophically and caused the frontiers of the Sahara to move northward. Like its contemporaries and predecessors, Rome was powerless in the face of major climatic shifts that affected vast areas of the world, many of them outside imperial frontiers.

Centralized preindustrial states were exceptionally vulnerable to megadroughts and other climate changes. Such developments as a series of weak monsoons or sudden climatic changes that caused floods to sweep away irrigation canals, followed by drought—as happened with Angkor—were beyond the capacity of rulers, however powerful, to survive. The states might implode, but out of the remaining fragments arose transformed societies, perhaps more dispersed, perhaps linked to new long-distance trade routes, but always lacking the industrial-scale infrastructure to manage increasing vulnerability and enhanced risk.

Over many centuries, preindustrial civilizations rose and fell with often habitual economic and political volatility. Almost without exception, they were vulnerable to climate change. When they adapted successfully, they did so at a local level, where able administrators could corral food supplies, close provincial frontiers, or deploy workers to build irrigation canals. The grandiloquent Egyptian nomarch Ankhtifi boasted on the walls of his tomb of his success in combatting drought in 2180 BCE. Even allowing for hyperbole, he had clearly realized a secret of successful adaptation: *local* measures are often far more effective than grandiose schemes that leave many people still at risk. His acolytes were constantly innovating and developing new methods of dealing with issues. Ultimately, this led to industrialization and, with it, ever more technology.

Modern technology gives us a huge advantage over our preindustrial forebears. Our technological abilities are such that we can land on and explore the moon, examine the deep trenches of the Pacific, and dabble in artificial intelligence. We're at the point where many people naively believe that technology will take care of climate change. Yes, it will help, as it did Roman roadbuilders and skippers of clipper ships, but the environmental cost has been, and will be, enormous. The ways in which we adapt to future climate challenges will indeed require massive investments in technological solutions, but they will have to be carbon neutral and self-sustaining. Such investments will be long-term and will require both huge expenditures and the political will to transform society and the ways in which we govern ourselves and do business. Technological innovations to control global climate change are probably within our reach, but the remit to achieve them is an enormous responsibility for future generations. As in the past, innovation brings responsibilities, but today this is the case on a scale unimaginable in the preindustrial world.

THE TIPPING POINT

For the first time, adapting to climate change has become as much a global issue as a local one. This is when the legacies of the past come to the forefront. The past has always been with us, offering encouragement, warning of pervasive dangers, and providing precedents for handling a crisis-ridden climatic future. Never have insights from ancient times been as important as they are today. For the first time we're causing massive climate change and interrupting the natural cycles of global climate. Rising carbon dioxide levels in the atmosphere, continuous and accelerating climbs in global temperatures, climbing sea levels destined to inundate enormous cities flourishing at or near sea level, chronic deforestation at human hands: ecological destruction surrounds us on every side in a world with over 7.6 billion people. Hundreds of millions of them live under threat of extreme weather events and drastic changes in great rivers like the Nile and the Mississippi resulting from anthropogenic climate change. The litany of potential climatic and ecological catastrophe surrounds us on every side, most of it the direct consequence of human activity. This is a far cry from the kinds of climatic adaptations that faced the Andeans, the Indus Civilization, medieval European farmers, and the Mughals of India. We're now at the moment when we have to confront unprecedented and utterly menacing global climate change.

Climatologists, ecologists, and highly respected scientists, as well as government agencies and international organizations, have repeatedly warned us of the crisis that lies ahead. But, like the Roman emperor Nero, we fiddle while our world is in danger of burning up and warming beyond recovery. There's an almost complete dearth of *global* leadership that looks not years or decades ahead but generations into the future and plans global strategies to create a safe world for our descendants. This is a truly global challenge that is unique in human experience and will exact a very high price indeed from us and from future generations. It's no exaggeration to imagine that the future of humankind is at stake.

The time for concerted action is at hand, and to ignore the lessons of the past, of hundreds of thousands of years of adapting to climate change is, to put it mildly, fatally shortsighted.

LESSONS FROM THE PAST

What are the lessons we've received from the past about adapting to climate change? They are brutally simple.

First, we are human beings, with the same behavioral qualities of every generation of *Homo sapiens*: brilliant qualities of forward thinking, of adept planning and cooperation, of intellectual reasoning and innovation. In planning adaptations to future climate change, we need to maximize these enduring qualities that will sustain us as we plan decisive adaptations for the future.

Second, we have developed a remarkable, and still rapidly improving, expertise at predicting climate change. The ancient societies described in these pages never had the benefits of scientific weather forecasting, satellite observations, the full array of proxies, and computer modeling that have revolutionized our knowledge of global climate and of the endless, unpredictable gavotte between the atmosphere and the oceans. The ancient Babylonians and others toyed with observations of the heavenly bodies as a way of predicting weather, without success, as did medieval European astronomers. Meteorologist Hubert Lamb remarked that weather predictions until the late nineteenth century were "church steeple meteorology," observations that involved gazing at cloud formations and other weather signs from elevated viewpoints.

Fully scientific meteorology is a product of the twentieth and twenty-first centuries. But much vital traditional climatic expertise still survives inconspicuously. Ancient Egyptian priests built

Nilometers to measure and predict the annual inundation. Early European mariners knew the telltale signs of approaching gales; Caribbean islanders and Maya astronomers could sometimes detect impending hurricanes; Pacific voyagers took advantage of 180-degree shifts in prevailing Polynesian trade winds to sail eastward during El Niños. We tend to ignore the remarkable predictive expertise of people still living close to the land and the ocean. Given the local impact of much climate change, this is a mistake. Much of this traditional knowledge, transmitted by word of mouth, still exists but needs to be collected before it is too late.

Third, in our preoccupation with global climate change, we forget that a great deal of adaptation to climate change, be it building seawalls or moving houses to higher ground, is a matter of *local* leadership and action. The local impacts of climate change emerge persistently in our narrative, and homegrown examples of successful adaptation abound. One notable example comes from Medmerry in southeastern England, where frequently inundated coastal land has been yielded to the ocean and turned into a nature reserve. Local adaptations, whatever the cost, are of paramount importance, even if they are the result of global climatic developments.

Fourth, we are social animals, which means that ties of family and wider kin, of community and nonprofit organizations with closely linked members, are a remarkable survival mechanism of fundamental importance in a world of unpleasant climatic crises. Such relationships have been part of human experience since the beginning. They are one of humankind's most powerful adaptive weapons; yet we consistently ignore their potential. Moreover, as social animals in a settled world, we tend to exploit both each other and the environment—whether in a bid to demonstrate our high(er) status over others or indeed to simply secure our survival where resources are finite and sometimes shaky. It is possible, in our opinion, to link many wars to conflicts over resources, whatever the claimed ideological or religious pretexts.

Fifth, we live in an industrialized world with exceptional infrastructures with vast potential for the future. But we tend to forget that countless people still live from harvest to harvest, often with uncertain water supplies and high vulnerability to hunger and drought. Relief aid in times of extreme drought and hunger is admirable but is not a lasting panacea. We were both struck by just how little attention is paid to the workings of traditional agriculture with its built-in, intimate knowledge of local environments. The Pueblo farmers of the Southwest, the Kekchi Maya of Belize, and the raised-field farmers of the Bolivian altiplano are examples of just how much we have to learn from traditional agriculture practiced successfully and out of the limelight for many centuries. A powerful legacy from the past, agricultural knowledge transmitted by word of mouth, is in danger of vanishing.

Lastly, preindustrial civilizations were remarkable for their volatility when confronted by climate crisis. Time and time again, even powerful leaders faltered in times of exigency, especially when drought and other climatic perturbations negated their perceived supernatural powers. Those who survived, either by taking action or by adapting thoughtfully to transformed circumstances, were decisive leaders, capable of thinking ahead and of bold action. Some of the anonymous leaders of Chimor on the Peruvian coast thought long-term. So did occasional Chinese emperors, their efforts often derailed by sclerotic bureaucracies. The past reminds us that the ultimate catalyst for long-term success in combatting climate change will be charismatic, authoritative leadership that transcends national interests and attacks climate change from a truly global perspective.

The starting point must be the reality that we are a global community of *Homo sapiens*, for our future depends on leadership that is not preoccupied with election cycles and other such trivia. The past reminds us that, for the first time, we are facing a truly global challenge that we have never encountered in three million years. This is because we are causing it and because there are so many more of us who will be impacted. We hope this book reveals the truth of past climate change and what life was really like through the testimony of archeologists and historians.

Will humankind survive? If the record of the past is any guide, we shall. But we will need to adapt, or perhaps be forced to adapt. There will be numerous challenges, almost certainly with violence and numerous casualties. The past reminds us that we are ingenious and innovative, capable of rising to trials that are far greater than those of ancient times. When we look back, as we now can in ways not previously dreamed possible, we can see what worked and what did not. But, perhaps most importantly, it is clear that we need to unite and cooperate as a species. We will survive, and one reason we will is because we've achieved an understanding of the complex relationship between humanity and the world's ever-changing climate. The past is not a foreign country—it's part of all of us, and it holds the keys to the future.

ACKNOWLEDGMENTS

Climate Chaos celebrates a major revolution in paleoclimatological research and new generations of multidisciplinary archeological and historical inquiry that are transforming our knowledge of how our forebears adapted to long- and short-term climate change. This has been a remarkable journey of exploration for both of us, drawing as it does on a very broad range of academic and nonacademic sources. Apart from broad-ranging climatological and historical topics, we've become literate in fascinating esoterica, among them Sumerian proverbs, Assyrian feasts, Maya water conservation, and the importance of mobility in human life. Among many other things, our research has encompassed the climatology of megadroughts and climatic proxies, sunspot maxima, subsistence agriculture, and even Mary Shelley's *Frankenstein*. But, above all, this book tries to answer a major question: What is the relevance of how ancient societies adapted to climate change to today's anthropogenically driven warming? Why is the past important for the climatic future? We believe that the past, its ever-changing climates, and its simple and much more complex societies have important lessons for the present and the future.

During the past fifteen years, Brian has written a series of books about ancient climate and human society revolving around such topics as El Niños, the Little Ice Age, and rising sea levels. All these books are now seriously out of date, a reflection of the dramatic advances in research in a few short years. Given the rapid advances in paleoclimatology and environmental history, a sequel is now overdue. This book attempts a reappraisal in a world where anthropogenic warming threatens our future as never before.

This is a complex historical puzzle that draws on a rapidly expanding, indeed frenziedly growing, academic literature, much of it highly technical, closely argued, and often contradictory. We were thankful to come across insightful books and papers that guided us through morasses of detail. Much of the literature behind this book is of a very high standard indeed, a particular reflection of the major advances in environmental history in recent years. The major practitioners are truly remarkable historical detectives. We've also benefitted from dozens of conversations with colleagues in many disciplines over the years, so many that it is impossible to thank them individually. Please forgive us if we offer a collective thank-you. Our paleoclimatological colleagues were especially generous with advice and encouragement, which we gratefully acknowledge.

Special thanks are due to Dagmar DeGroot, Kyle Harper, Charles Higham, Lisa Lucero, Michael Mann, Michael McCormick, William Marquardt, Paul Mayewski, George Michaels, Vernon Scarborough, Sam White, and the late Professor Grahame Clark, who introduced Brian to climate change and environmental archeology.

Our agent, Susan Rabiner, has been, as always, a great inspiration and a true friend to us both. Ben Adams at Public Affairs has always encouraged us, and we are grateful for his shrewd insights. Mitch Allen read through the entire manuscript with his usual perceptive skill and improved it dramatically. Shelly Lowenkopf was a tower of strength and made a huge difference to the narrative.

Lastly, a word of thanks to our families. Brian could not have written the book without Lesley and Ana's constant support, to say nothing of his feline colleague, Atticus Catticus the Moose, who prefers sunbathing to writing. Nadia thanks Matthew and Jacob for their good humor and encouragement—with added appreciation to Matin, Katja, Chiara, and Alex.

NOTES

The literature on ancient climate and paleoclimatology is enormous and growing daily, to the point that it is becoming harder and harder to stay abreast of the latest research. The notes that follow point you to major sources with comprehensive biographies that can guide you into the perils of more specialized literature. These notes have no ambitions to be comprehensive.

Prolegomenon: Before We Begin

1. S. George Philander, *Is the Temperature Rising? The Uncertain Science of Global Warming* (Princeton, NJ: Princeton University Press, 1998).

1. A Frozen World (c. 30,000 to c. 15,000 Years Ago)

1. John F. Hoffecker, *A Prehistory of the North* (New Brunswick, NJ: Rutgers University Press, 2005).

2. Brian Fagan, ed., *The Complete Ice Age* (London and New York: Thames & Hudson, 2009), is a collection of popular essays by experts. For Ice Age temperatures, see Jessica Tierney et al., "Glacial Cooling and Climate Sensitivity Revisited," *Nature* 584 (2020): 569–573. doi: 10.1038/s41586-020-2617-x.

3. Brian Fagan, *Cro-Magnon: How the Ice Age Gave Birth to the First Modern Humans* (New York: Bloomsbury Press, 2010).

4. Ian Gilligan, *Climate, Clothing, and Agriculture in Prehistory: Linking Evidence, Causes, and Effects* (Cambridge: Cambridge University Press, 2018), is a definitive and thoughtful analysis of this subject.

5. Charles Darwin, *Charles Darwin's "Beagle" Diary*, ed. Richard Darwin Keynes (Cambridge: Cambridge University Press, 1988), 134.

6. Paul H. Barrett and R. B. Freeman, *Journal of Researches: The Works of Charles Darwin* (New York: New York University Press, 1987), pt. 3, 2:120.

7. John F. Hoffecker, *Desolate Landscapes: Ice-Age Settlement in Eastern Europe* (New Brunswick, NJ: Rutgers University Press, 2002), chap. 5.

8. Fagan, *Cro-Magnon*, 159–163.

9. Hoffecker, *Prehistory of the North*, chaps. 5 and 6.

10. Hoffecker, *Prehistory of the North*, chaps. 5 and 6.

11. Jean Combier and Anta Montet-White, eds., *Solutré 1968–1998*. Memoir XXX (Paris: Société Préhistorique Française, 2002).

12. Olga Soffer, *The Upper Palaeolithic of the Eastern European Plain* (New York: Academic Press, 1985).

2. After the Ice (Before 15,000 Years Ago to c. 6000 BCE)

1. A Greek philosopher, Diogenes of Sinope (386–354 BCE) is said to have traveled inland from the town of Rhapta somewhere in present-day southern Tanzania for twenty-five days. He named the Rwenzoris "the Mountains of the Moon" and believed they were the source of the Nile. The geographer Marinus of Tyre (c. 70–130 CE) recorded Diogenes's travels, which provided the underpinnings of Ptolemy's *Geography*. Unfortunately, Marinus's geographical treatise is lost. Later Arab travelers duly called the legendary peaks Jibbel el Kumri ("Mountains of the Moon" in Arabic). In 1889, explorer Henry Morton Stanley of "Dr. Livingstone, I presume" fame definitively located the mountains on a map. No European traveler had seen them earlier because they were usually shrouded in clouds.

2. Margaret S. Jackson et al., "High-Latitude Warming Initiated the Onset of the Fast Deglaciation in the Tropics," *Science Advances* 5 (12) (2019). doi: 10.1126/sciadv.aaw2610.

3. Steven Mithen, *After the Ice: A Global Human History, 20,000–5000 BC* (Cambridge, MA: Harvard University Press, 2006), is an authoritative and provocative summary.

4. Vincent Gaffney et al., *Europe's Lost World: The Rediscovery of Doggerland* (York: Council for British Archaeology, 2009).

5. The literature on the first settlement of the Americas is enormous and riddled with controversy. See David Meltzer, *First Peoples in a New World: Colonizing Ice Age America* (Berkeley: University of California Press, 2008). See also the same author's *The Great Paleolithic War: How Science Forged an Understanding of America's Ice Age Past* (Chicago: University of Chicago Press, 2015).

6. Again, the literature is vast and contradictory. A useful summary: Graeme Barker, *The Agricultural Revolution in Prehistory* (New York: Oxford University Press, 2006).

7. Bruce G. Trigger, *Gordon Childe: Revolutions in Archaeology* (New York: Columbia University Press, 1980), is the best source on Childe's ideas and work.

8. William Ruddiman, *Plows, Plagues, and Petroleum: How Humans Took Control of Climate* (Princeton, NJ: Princeton University Press, 2016).

9. Egyptologist James Henry Breasted coined the term "Fertile Crescent" in popular books a century ago. It refers to a giant semicircle, open to the south, that arches from the southeast corner of the Mediterranean, north through Syria, part of Turkey, and the Iranian highlands, then south to the Persian Gulf. Breasted likened it to a "desert bay." The Fertile Crescent is purely a convenient label with no rigid definition, but it has stood the test of time.

10. Klaus Schmidt, *Göbekli Tepe: A Stone Age Sanctuary in South-eastern Turkey* (London: ArchaeNova, 2012).

11. Andrew T. Moore et al., *Village on the Euphrates* (New York: Oxford University Press, 2000).

12. By any standards, Çatalhöyük is a truly remarkable long-term archaeological project carried out by international teams of excavators and researchers. The literature is growing rapidly. The best starting point for general readers: Ian Hodder, *The Leopard's Tale* (London and New York: Thames & Hudson, 2011). On a more technical level, the same author's edited *Religion in the Emergence of Civilization: Çatalhöyük as a Case Study* (Cambridge: Cambridge University Press, 2010) is a fascinating excursion into the archaeology of the intangible.

3. Megadrought (c. 5500 BCE to 651 CE)

1. Nicola Crusemann et al., eds., *Uruk: First City of the Ancient World* (Los Angeles: J. Paul Getty Museum, 2019).

2. Monica Smith, *Cities: The First 6,000 Years* (New York, Penguin, 2019).

3. T. J. Wilkinson, *Archaeological Landscapes of the Near East* (Tucson: University of Arizona Press, 2003).

4. Samuel Kramer, *The Sumerians* (Chicago: University of Chicago Press, 1963), 240.

5. Mario Liverani, *The Ancient Near East: History, Society and Economy* (Abingdon, UK: Routledge, 2014).

6. Kramer, *The Sumerians*, 190.

7. William H. Stiebing and Susan L. Helft, *Ancient Near Eastern History and Culture*, 3rd ed. (Abingdon, UK: Routledge, 2017). See also Benjamin Foster, *The Age of Agade: Inventing Empire in Ancient Mesopotamia* (Abingdon, UK: Routledge, 2016).

8. J. S. Cooper, "Reconstructing History from Ancient Inscriptions: The Lagash-Umma Border Conflict," *Sources and Monographs on the Ancient Near East* 2, no. 1 (1983): 47–54.

9. Marc Van De Mieroop, *A History of the Ancient Near East ca. 3000–323 BC*, 2nd ed. (New York: Blackwell, 2006). See also Foster, *The Age of Agade*.

10. This passage, indeed, this entire chapter, owes much to Harvey Weiss's admirable account of climate change and the Akkadian collapse. Harvey Weiss, "4.2 ka BP Megadrought and the Akkadian Collapse," in *Megadrought and Collapse: From Early Agriculture to Angkor*, ed. Harvey Weiss (New York: Oxford University Press, 2017), 93–159. A growing literature surrounds the drought and its causes. See Heidi M. Cullen et al., "Impact of the North Atlantic Oscillation on Middle Eastern Climate and Streamflow," *Climatic Change* 55 (2002): 315–338. See also Martin H. Visbeck et al., "The North Atlantic Oscillation: Past, Present, and Future," *Proceedings of the National Academy of Sciences* 98, no. 23 (2001): 12876–12877.

11. Weiss, "4.2 ka BP Megadrought and the Akkadian Collapse," 135–159, has an invaluable list of proxy sites with references.

12. M. Charles, H. Pessin, and M. M. Hald, "Tolerating Change at Late Chalcolithic Tell Brak: Responses of an Early Urban Society to an Uncertain Climate," *Environmental Archaeology* 15, no. 2 (2010): 183–198.

13. Charles, Pessin, and Hald, "Tolerating Change at Late Chalcolithic Tell Brak," 183–198.

14. W. Sallaberger, "Die Amurriter-Mauer in Mesopotamien: der älteste historische Grenzwall gegen Nomaden vor 4000 Jahren," in *Mauern als Grenzen*, ed. A. Nunn (Mainz: Phillipp von Zabern, 2009), 27–38.

15. J. A. Black et al., *The Literature of Ancient Sumer* (New York: Oxford University Press, 2004), 128–131.

16. The feast is described in a royal inscription from Kalhu. Van De Mieroop, *A History of the Ancient Near East*, 234.

17. Kuna Ba: Ashish Sinha et al., "Role of Climate in the Rise and Fall of the Neo-Assyrian Empire," *Science Advances* 5, no. 11 (2019). doi: 10.1126/sciadv.aax6656.

18. Nathan J. Wright et al., "Woodland Modification in Bronze and Iron Age Central Anatolia: An Anthracological Signature for the Hittite State?" *Journal of Archaeological Science* 55 (2015): 219–230.

19. Touraj Daryaee, *Sasanian Persia: The Rise and Fall of an Empire*. Rpt. ed. (New York: I. B. Tauris, 2013). See also Eberhard Sauer, ed., *Sasanian Persia: Between Rome and the Steppes of Eurasia* (Edinburgh: Edinburgh University Press, 2019).

20. Fagan, *Cro-Magnon*, 146–152.

4. Nile and Indus (3100 to c. 1700 BCE)

1. Herodotus, *The Histories*, trans. Robin Waterfield (Oxford: Oxford University Press, 1998), bk. 2, line 111, 136.

2. J. Donald Hughes, "Sustainable Agriculture in Ancient Egypt," *Agricultural History* 66, no. 2 (1992): 13.

3. Barry Kemp, *Ancient Egypt: The Anatomy of a Civilization*, 3rd ed. (Abingdon, UK: Routledge, 2018), is an admirable guide to ancient Egyptian civilization.

4. I. E. S. Edwards, *The Pyramids of Egypt* (Baltimore: Pelican, 1985), 12.

5. Mark Lehner, *The Complete Pyramids* (London: Thames & Hudson, 1997). See also Miroslav Verner, *The Pyramids*. Rev. ed. (Cairo: American University in Cairo Press, 2021).

6. The length of Pepi II's reign is disputed and may have been as short as sixty-four years, still an impressive reign by pharaonic standards.

7. The role of climate change in the collapse of the Old Kingdom is a controversial issue in Egyptology. A useful summary of the arguments can be found in Ellen Morris, "Ancient Egyptian Exceptionalism: Fragility, Flexibility and the Art of Not Collapsing," in *The Evolution of Fragility: Setting the Terms*, ed. Norman Yoffee (Cambridge, UK: McDonald Institute for Archaeological Research, 2019), 61–88.

8. *The Admonitions of Ipuwer*, thought to date to the Middle Kingdom, is an incomplete literary work, preserved in a papyrus dating to about 1250 BCE, but the text is from much earlier. It is the earliest-known treatise on political ethics. Ipuwer argued that a good pharaoh should control his officials and carry out the will of the gods. Quotes from Barbara Bell, "Climate and the History of Egypt: The Middle Kingdom," *American Journal of Archaeology* 79 (1975): 261.

9. Barbara Bell, "The Dark Ages in Ancient History, I: The First Dark Age in Egypt," *American Journal of Archaeology* 75 (1971): 9.

10. A general description of the Indus civilization: Andrew Robinson, *The Indus: Lost Civilizations* (London: Reaktion, 2021). See also Robin Coningham and Ruth Young,

From the Indus to Ashoka: Archaeologies of South Asia (Cambridge: Cambridge University Press, 2015).

11. Ashish Sinha et al, "Trends and Oscillations in the Indian Summer Monsoon Rainfall over the Past Two Millennia," *Nature Communications* 6, no. 6309 (2015); Peter B. deMenocal, "Cultural Responses to Climate Change During the Late Holocene," *Science* 292, no. 5517 (1976): 667–673. See also Alena Giesche et al., "Indian Winter and Summer Monsoon Strength over the 4.2 ka BP Event in Foraminifer Isotope Records from the Indus River Delta in the Arabian Sea," *Climate of the Past* 15, no. 1 (2019): 73. doi: 10.5194/cp-15-73-2019.

12. Gayatri Kathayat et al., "The Indian Monsoon Variability and Civilization Changes in the Indian Subcontinent," *Science Advances* 3 (2017): e1701296.

13. Mortimer Wheeler, *The Indus Civilization*, 3rd ed. (Cambridge: Cambridge University Press, 1968), 44.

14. A basic source: Cameron A. Petrie, "Diversity, Variability, Adaptation, and 'Fragility' in the Indus Civilization," in Yoffee, *Evolution of Fragility*, 109–134.

15. C. A. Petrie and J. Bates, "'Multi-cropping', Intercropping and Adaptation to Variable Environments in Indus South Asia," *Journal of World Prehistory* 30 (2017): 81–130, is a comprehensive treatment of Indus agriculture.

5. The Fall of Rome (c. 200 BCE to the Eighth Century CE)

1. Neither of us are Romanists, so this chapter relies heavily on Kyle Harper's closely argued synthesis: *The Fate of Rome: Climate, Disease, and the End of an Empire* (Princeton, NJ: Princeton University Press, 2017). Harper marshals a broad array of sources to discuss the central roles of climate change and pandemics in the empire's prolonged implosion. At times controversial and provocative, this remarkable book guides the reader skillfully through the intricacies of the subject. We have, of course, glossed over the numerous controversies and points of disagreement in this brief summary. Harper includes a comprehensive bibliography. See also Rebecca Storey and Glenn R. Storey, *Rome and the Classic Maya* (Abingdon, UK: Routledge, 2017).

2. An overview of Roman climate appears in Kyle Harper and M. McCormick, "Reconstructing the Roman Climate," in *The Science of Roman History*, ed. W. Scheidel (Princeton, NJ: Princeton University Press, in preparation). Also, a major synthesis: Michael McCormick et al., "Climate Change During and After the Roman Empire: Reconstructing the Past from Scientific and Historical Evidence," *Journal of Interdisciplinary History* 43, no. 2 (2012): 169–220. The Okmok II eruption: Joseph R. McConnell et al., "Extreme Climate After Massive Eruption of Alaska's Okmok Volcano in 43 BCE and Effects on the Late Roman Republic and Ptolomaic Kingdom," *Proceedings of the National Academy of Sciences* 117, no. 27 (July 7, 2020): 15443–15449. doi: 10.1073/pnas.2002722117.

3. *Foggaras* are gently sloping underground channels or tunnels that tap aquifers or deep wells to irrigate farming land. They are known in Iran as *qanats* and were widely used in the Middle East and North Africa for many centuries. Basically, they are underground aqueducts.

4. Quote and source for the paragraph: Harper, *Fate of Rome*, 53–54.

5. Harper, *Fate of Rome*, 54.

6. Sigwells: Richard Tabor, *Cadbury Castle: The Hillfort and Landscapes* (Stroud, UK: History Press, 2008), 130–142. Catsgore: R. Leech, *Excavations at Catsgore, 1970–1973* (Bristol, UK: Western Archaeological Trust, 1982).

7. Harper, *Fate of Rome*, 57.

8. Harper, *Fate of Rome*, 57–58.

9. A *modius* (pl. *modii*) is equivalent to a peck, or roughly nine liters of dry goods.

10. These passages based on Harper, *Fate of Rome*, 92–98. A summary of Indian Ocean seafaring and trade appears in Brian Fagan, *Beyond the Blue Horizon: How the Earliest Mariners Unlocked the Secrets of the Oceans* (New York: Bloomsbury Press, 2012), chaps. 7 to 9.

11. Hui-Yuan Yeh et al., "Early Evidence for Travel with Infectious Diseases Along the Silk Road: Intestinal Parasites from 2000-Year-Old Personal Hygiene Sticks in a Latrine at Xuanquanzhi Relay Station in China," *Journal of Archaeological Science: Reports* 9 (2016): 758–764.

12. William H. McNeill, *Plagues and Peoples* (New York: Doubleday, 1976), and Harper, *Fate of Rome*, chap. 3, cover the Antonine Plague.

13. Cyprian (c. 200–258 CE) was of Berber ancestry but became bishop of Carthage and a well-known early Christian writer. The plague he described was named after him. Quote from Harper, *Fate of Rome*, 130.

14. For general descriptions, see Lucy Grig and Gavin Kelly, eds., *Two Romes: Rome and Constantinople in Late Antiquity* (Oxford: Oxford University Press, 2012).

15. Harper, *Fate of Rome*, 185.

16. M. Finné et al., "Climate in the Eastern Mediterranean, and Adjacent Regions During the Past 6000 Years—a Review," *Journal of Archaeological Science* 38 (2011): 3153–3173.

17. E. Cook, "Megadroughts, ENSO, and the Invasion of Late-Roman Europe by the Huns and Avars," in *The Ancient Mediterranean Environment Between Science and History*, ed. William Harris (Leiden: Brill, 2013), 89–102. See also Q-Bin Zhang et al., "A 2,326-Year Tree-ring Record of Climate Variability on the Northeastern Qinghai-Tibetan Plateau," *Geophysical Research Letters* 30, no. 14 (2003). doi: 10.1029/2003GL017425.

18. Quoted from Harper, *Fate of Rome*, 192. Ammianus Marcellinus (354–378 CE) was a soldier and the last great Roman historian. His major work was *Res gestae*, a thirty-one-volume history that takes off where Tacitus ended. The first thirteen volumes are lost.

19. Described by Harper, *Fate of Rome*, 199–200.

20. We have relied on Harper, *Fate of Rome*, chap. 6, when summarizing the Justinian Plague. However, much needs to be learned about the local impacts of the plague and accompanying mortality rates—also about the history of *Yersinia pestis*. See also William Rosen, *Justinian's Flea* (New York: Penguin Books, 2008).

21. John of Ephesus (c. 507–588 CE) was a leader of the Syriac Orthodox Church and a historian. The third part of his *Ecclesiastical History* covers the Justinian Plague, which he witnessed firsthand. He considered it a sign of divine wrath. Quote from Harper, *Fate of Rome*, 227.

22. Stuart J. Borsch, "Environment and Population: The Collapse of Large Irrigation Systems Reconsidered," *Comparative Studies in Society and History* 46, no. 3 (2004): 451–468, and other papers by the same author.

23. The Roman statesman Cassiodorus (c. 485–585 CE) was also a respected scholar and writer. He founded the Vivarium monastery, devoted to reading and copying manuscripts, at his estate on the Ionian Sea.

24. Edward Gibbon (1737–1794) was a historian and member of Parliament and the author of an immortal work, *The History of the Decline and Fall of the Roman Empire*, published in six volumes between 1776 and 1788. Edward Gibbon and David P. Womersley, *History of the Decline and Fall of the Roman Empire*, 3 vols. (London: Penguin Press, 1994).

6. The Maya Transformation (c. 1000 BCE to the Fifteenth Century CE)

1. The term "Mesoamerica" is used in scholarly circles to refer to the area of Central America where preindustrial civilizations developed (encompassing today's central Mexico, Belize, Guatemala, El Salvador, Honduras, Nicaragua, and northern Costa Rica).

2. The terminology of Maya civilization has at its core Classic Maya civilization, which lasted from about 250 to 900 CE. We use this terminology here for convenience, but it does, of course, mask considerable cultural diversity.

3. For a thoughtful description of the lowlands, which we have drawn on here, see B. J. Turner II and Jeremy A. Sabloff, "Classic Period Collapse of the Central Maya Lowlands: Insights About Human-Environment Relationships for Sustainability," *Proceedings of the National Academy of Sciences* 109, no. 35 (2012): 13908–13914.

4. The classic popular account of ancient Maya civilization is Michael Coe and Stephen Houston, *The Maya*, 9th ed. (London and New York: Thames & Hudson, 2015). Linda Schele and David Freidel's *A Forest of Kings* (New York, William Morrow, 1990), is a vivid popular account of Maya kingship, which is now somewhat outdated.

5. Richard R. Wilk, "Dry-Season Agriculture Among the Kekchi Maya and Its Implications for Prehistory," in *Prehistoric Lowland Maya Environment and Subsistence Economy*, ed. Mary Pohl (Cambridge, MA: Peabody Museum of Archaeology and Ethnology, Harvard University, 1985), 47–57. See also the same author's *Household Ecology: Economic Change and Domestic Life Among the Kekchi Maya of Belize*. Arizona Studies in Human Ecology (Tucson: University of Arizona Press, 1991).

6. B. L. Turner II, "The Rise and Fall of Maya Population and Agriculture: The Malthusian Perspective Reconsidered," in *Hunger and History: Food Shortages, Poverty, and Deprivation*, ed. L. Newman (Cambridge: Cambridge University Press, 1990), 178–211.

7. Robert J. Oglesby et al., "Collapse of the Maya: Could Deforestation Have Contributed?" *Papers in the Earth and Atmospheric Sciences* 469 (2010). http://digitalcommons.unl.edu/geosciencefacpub/469.

8. The literature on the Classic Maya collapse is enormous. For general summaries, see T. Patrick Culbert, ed., *The Classic Maya Collapse* (Albuquerque: University of New Mexico Press, 1973), now somewhat outdated, and D. Webster, *The Fall of the Ancient Maya* (London and New York: Thames & Hudson, 2002). A useful analysis, which we

relied on heavily here, is Turner and Sabloff, "Classic Period Collapse of the Central Maya Lowlands."

9. David Hodell, M. Brenner, and J. H. Curtis, "Terminal Classic Drought in the Northern Maya Lowlands Inferred from Multiple Sediment Cores in Lake Chichancanab (Mexico)," *Quaternary Science Reviews* 24 (2005): 1413–1427.

10. Douglas Kennett and David A. Hodell, "AD 750–100 Climate Change and Critical Transitions in Classic Maya Sociopolitical Networks," in *Megadrought and Collapse: From Early Agriculture to Angkor*, ed. Harvey Weiss (New York: Oxford University Press, 2017), 204–230. See also Douglas Kennett et al., "Development and Disintegration of Maya Political Systems in Response to Climate Change," *Science* 338 (2012): 788–791.

11. Copán: William L. Fash and Ricardo Agurcia Fasquelle, "Contributions and Controversies in the Archaeology and History of Copán," in *Copán: The History of an Ancient Maya Kingdom*, ed. E. Wyllys Andrews and William L. Fash (Santa Fe, NM: School of American Research Press, 2005), 3–32. See also William L. Fash, E. Wyllys Andrews, and T. Kam Manahan, "Political Decentralization, Dynastic Collapse, and the Early Postclassic in the Urban Center of Copán, Honduras," in *The Terminal Classic in the Maya Lowlands: Collapse, Transition, and Transformation*, ed. Arthur A. Demarest, Prudence M. Rice, and Don S. Rice (Boulder: University Press of Colorado, 2005), 260–287.

12. Arthur Demarest, *Ancient Maya: Rise and Fall of a Rainforest Civilization* (Cambridge: Cambridge University Press, 2004).

13. Jeremy A. Sabloff, "It Depends on How You Look at Things: New Perspectives on the Postclassic Period in the Northern Maya Lowlands," *Proceedings of the American Philosophical Society* 109 (2007): 11–25. See also Marilyn A. Masson, "Maya Collapse Cycles," *Proceedings of the National Academy of Sciences* 109, no. 45 (2012): 18237–18238.

14. Marilyn A. Masson and Carlos Peraza Lope, *Kukulkan's Realm: Urban Life at Mayapan* (Boulder: University of Colorado Press, 2014), 5.

7. Gods and El Niños
(c. 3000 BCE to the Fifteenth Century CE)

1. L. G. Thompson et al., "A 1500-Year Record of Climate Variability Recorded in Ice Cores from the Tropical Quelccaya Ice Cap," *Science* 229 (1985): 971–973.

2. Michael Moseley, *The Inca and Their Ancestors*, 2nd ed. (London and New York: Thames & Hudson, 2001), is a widely quoted synthesis.

3. Ruth Shady and Christopher Kleihege, *Caral: First Civilization in the Americas*. Bilingual ed. (Chicago: CK Photo, 2010).

4. Moche: Apart from Moseley, *The Inca and Their Ancestors*, see Jeffrey Quilter, *The Ancient Central Andes* (Abingdon, UK: Routledge, 2013).

5. Walter Alva and Christopher Donnan, *Royal Tombs of Sipán* (Los Angeles: Fowler Museum of Cultural History, 1989). An update: Nadia Durrani, "Gold Fever: The Tombs of the Lords of Sipan," *Current World Archaeology* 35 (2009): 18–30.

6. L. G. Thompson et al., "Annually Resolved Ice Core Records of Tropical Climate Variability over the Past 1800 Years," *Science* 229 (2013): 945–950.

7. Brian Fagan, *Floods, Famines, and Emperors: El Niño and the Fate of Civilizations*. Rev. ed. (New York: Basic Books, 2009), chap. 7, has a description for general readers.

8. Michael Moseley and Kent C. Day, eds., *Chan Chan: Andean Desert City* (Albuquerque: University of New Mexico Press, 1982).

9. Brian Fagan, *The Great Warming* (New York: Bloomsbury Press, 2008), chap. 9, has a general description.

10. Charles R. Ortloff, "Canal Builders of Pre-Inca Peru," *Scientific American* 359, no. 6 (1988): 100–107.

11. Tom D. Dillehay and Alan L. Kolata, "Long-Term Human Response to Uncertain Environmental Conditions in the Andes," *Proceedings of the National Academy of Sciences* 101, no. 2: 4325–4330.

12. Alan L. Kolata, *The Tiwanaku: Portrait of an Andean Civilization* (Cambridge, MA: Blackwell, 1993). Two edited volumes are detailed monographs: Alan L. Kolata, ed., *Tiwanaku and Its Hinterland: Archaeology and Paleoecology of an Andean Civilization*, vol. 1: *Agroecology* and vol. 2: *Urban and Rural Archaeology* (Washington, DC: Smithsonian Institution, 1996 and 2003).

13. Charles Stanish et al., "Tiwanaku Trade Patterns in Southern Peru," *Journal of Anthropological Archaeology* 29 (2010): 524–532.

14. This section relies heavily on Lonnie G. Thompson and Alan L. Kolata, "Twelfth Century A.D.: Climate, Environment, and the Tiwanaku State," in *Megadrought and Collapse: From Early Agriculture to Angkor*, ed. Harvey Weiss (New York: Oxford University Press, 2017), 231–246.

15. R. A. Covey, "Multiregional Perspectives on the Archaeology of the Andes During the Late Intermediate Period (c. A.D. 1000–1400)," *Journal of Archaeological Research* 16 (2008): 287–338.

16. E. Arkush, *Hillforts of the Ancient Andes: Colla Warfare, Society, and Landscape* (Gainesville: University Press of Florida, 2011). See also E. Arkush and T. Tung, "Patterns of War in the Andes from the Archaic to the Late Horizon: Insights from Settlement Patterns and Cranial Trauma," *Journal of Archaeological Research* 219, no. 4 (2013): 307–369; Alan L. Kolata, C. Stanish, and O. Rivera, eds., *The Technology and Organization of Agricultural Production in the Tiwanaku State* (Pittsburgh, PA: Pittsburgh Foundation, 1987).

17. Clark L. Erickson, "Applications of Prehistoric Andean Technology: Experiments in Raised Field Agriculture, Huatta, Lake Titicaca, 1981–2," in *Prehistoric Intensive Agriculture in the Tropics*, ed. I. S. Farrington. International Series 232 (Oxford: British Archaeological Reports, 1985), 209–232. An invaluable paper on traditional farming in this region is Clark Erickson, "Neo-environmental Determinism and Agrarian 'Collapse' in Andean Prehistory," *Antiquity* 73 (1999): 634–642.

8. Chaco and Cahokia (c. 800 to 1350 CE)

1. Brian Fagan, *Before California: An Archaeologist Looks at Our Earliest Inhabitants* (Lanham, MD: Rowman & Littlefield, 2003); Jeanne Arnold and Michael Walsh, *California's Ancient Past: From the Pacific to the Range of Light* (Washington, DC: Society for American Archaeology, 2011).

2. Lynn H. Gamble, *First Coastal Californians* (Santa Fe, NM: School for Advanced Research, 2015), is an admirable account for general readers.

3. Douglas J. Kennett and James P. Kennett, "Competitive and Cooperative Responses to Climatic Instability in Coastal Southern California," *American Antiquity* 65 (2000): 379–395. See also Douglas J. Kennett, *The Island Chumash: Behavioral Ecology of a Maritime Society* (Berkeley: University of California Press, 2005).

4. Lynn H. Gamble, *The Chumash World at European Contact* (Berkeley: University of California Press, 2011).

5. Frances Joan Mathien, *Culture and Ecology of Chaco Canyon and the San Juan Basin* (Santa Fe, NM: National Park Service, 2005). See also Gwinn Vivian, *Chacoan Prehistory of the San Juan Basin* (New York: Academic Press, 1990).

6. An account of Chaco for a general audience is Brian Fagan, *Chaco Canyon: Archaeologists Explore the Lives of an Ancient Society* (New York: Oxford University Press, 2005). Essays on recent research in the canyon appear in Jeffrey J. Clark and Barbara J. Mills, eds., "Chacoan Archaeology at the 21st Century," *Archaeology Southwest* 32, nos. 2–3 (2018).

7. Jill E. Neitzel, *Pueblo Bonito: Center of the Chacoan World* (Washington, DC: Smithsonian Books, 2003). See also Timothy R. Pauketat, "Fragile Cahokian and Chacoan Orders and Infrastructures," in *The Evolution of Fragility: Setting the Terms*, ed. Norman Yoffee (Cambridge, UK: McDonald Institute for Archaeological Research, 2019), 89–108.

8. Vernon Scarborough et al., "Water Uncertainty, Ritual Predictability and Agricultural Canals at Chaco Canyon, New Mexico," *Antiquity* 92, no. 364 (August 2018): 870–889.

9. Douglas L. Kennett et al., "Archaeogenomic Evidence Reveals Prehistoric Patrilineal Dynasty," *Nature Communications* 8, no. 14115 (2017). doi: 10.1038/ncomms14115.

10. This section is based on David W. Stahle et al., "Thirteenth Century A.D.: Implications of Seasonal and Annual Moisture Reconstructions for Mesa Verde, Colorado," in Weiss, *Megadrought and Collapse*, 246–274. See also Mark Varien et al., "Historical Ecology in the Mesa Verde Region: Results from the Village Ecodynamics Project," *American Antiquity* 72 (2007): 273–299.

11. Cahokia: The literature is vast. See Timothy R. Pauketat, *Cahokia: Ancient America's Great City on the Mississippi* (New York: Viking Penguin, 2009), and the same author's *Ancient Cahokia and the Mississippians* (Cambridge: Cambridge University Press, 2004). See also Timothy R. Pauketat and Susan Alt, eds., *Medieval Mississippians: The Cahokian World* (Santa Fe, NM: School of Advanced Research, 2015); Pauketat, "Fragile Cahokian and Chacoan Orders and Infrastructures," 89–108.

12. A. J. White et al., "Fecal Stanols Show Simultaneous Flooding and Seasonal Precipitation Change Correlate with Cahokia's Population Decline," *Proceedings of the National Academy of Sciences* 116, no. 12 (2019): 5461–5466.

13. Samuel E. Munoz et al., "Cahokia's Emergence and Decline Coincided with Shifts of Flood Frequency on the Mississippi River," *Proceedings of the National Academy of Sciences* 112, no. 20 (2015): 6319–6327. See also Timothy R. Pauketat, "When the Rains Stopped: Evapotranspiration and Ontology at Ancient Cahokia," *Journal of Anthropological Research* 76, no. 4 (2020): 410–438.

14. A. J. White et al., "After Cahokia: Indigenous Repopulation and Depopulation of the Horseshoe Lake Watershed AD 1400–1900," *American Antiquity* 85, no. 2 (April 2020): 263–278.

9. The Disappeared Megacity (802 to 1430 CE)

1. For a general account of Khmer civilization, see Charles Higham, *The Civilization of Angkor* (London: Cassel, 2002), or Michael D. Coe, *Angkor and the Khmer Civilization* (London and New York: Thames & Hudson, 2005). See also Roland Fletcher et al., "Angkor Wat: An Introduction," *Antiquity* 89, no. 348 (2015): 1388–1401.

2. For a popular account of the latest research, see Brian Fagan and Nadia Durrani, "The Secrets of Angkor Wat," *Current World Archaeology* 7, no. 5 (2016):14–21.

3. LiDAR at Angkor: Damian Evans et al., "Uncovering Archaeological Landscapes at Angkor Using Lidar," *Proceedings of the National Academy of Sciences* 110 (2013): 12595–12600.

4. Roland Fletcher et al., "The Water Management Network of Angkor, Cambodia," *Antiquity* 82 (2008): 658–670.

5. The remainder of this chapter is based on Roland Fletcher et al., "Fourteenth to Sixteenth Centuries AD: The Case of Angkor and Monsoon Extremes in Mainland Southeast Asia," in *Megadrought and Collapse: From Early Agriculture to Angkor*, ed. Harvey Weiss (New York: Oxford University Press, 2017), 275–313; quote from 279.

6. P. D. Clift and R. A. Plumb, *The Asian Monsoon: Causes, History, and Effects* (Cambridge: Cambridge University Press, 2008).

7. A summary of this complex process of deterioration lies in Fletcher, "Fourteenth to Sixteenth Centuries AD," 292–304.

8. B. M. Buckley et al., "Climate as a Contributing Factor in the Demise of Angkor, Cambodia," *Proceedings of the National Academy of Sciences* 107 (2010): 6748–6752. See also B. M. Buckley et al., "Central Vietnam Climate over the Past Five Centuries from Cypress Tree Rings," *Climate Dynamics Heidelberg* 48, nos. 11–12 (2017): 3707–3708.

9. Dandak: A. Sinha et al., "A Global Context for Megadroughts in Monsoon Asia During the Past Millennium," *Quaternary Science Reviews* 30 (2010): 47–62. Wanxiang speleothems: R.-H Zhang et al., "A Test of Climate, Sun, and Culture Relationships from an 1810-Year Chinese Cave Record," *Science* 322 (2008): 940–942.

10. R. A. E. Coningham and M. J. Manson, "The Early Empires of South Asia," in *Great Empires of the Ancient World*, ed. T. Harrison (London and New York: Thames & Hudson, 2009), 226–249.

11. De Silva, K. M., *A History of Sri Lanka* (New Delhi: Penguin Books, 2005).

12. R. A. E. Coningham, *Anuradhapura: The British–Sri Lankan Excavations at Anuradhapura Salgaha Watta*. 3 vols. (Oxford, UK: Archaeopress for the Society for South Asian Studies, 1999, 2006, 2013).

13. Lisa J. Lucero, Roland Fletcher, and Robin Coningham, "From 'Collapse' to Urban Diaspora: The Transformation of Low-Density, Dispersed Agrarian Urbanism," *Antiquity* 89, no. 337 (2015): 1139–1154.

14. Mike Davis, *Late Victorian Holocausts: El Niño Famines and the Making of the Third World* (Brooklyn, NY: Verso Books, 2001).

15. Frederick Williams, *The Life and Letters of Samuel Wells Williams, MD: Missionary, Diplomatist, Sinologue* (New York: G. P. Putnam's Sons, Knickerbocker Press, 1889), 432.

10. Africa's Reach (First Century BCE to 1450 CE)

1. Mike Davis, *Late Victorian Holocausts: El Niño Famines and the Making of the Third World* (Brooklyn, NY: Verso Books, 2001), 201.

2. Davis, *Late Victorian Holocausts*, 201.

3. Brian Fagan, *Floods, Famines, and Emperors: El Niño and the Fate of Civilizations*. Rev. ed. (New York: Basic Books, 2009), 16. Abu Zayd Al-Sirafi was a seafarer. In about 916 CE, he wrote *Accounts of China and India*, trans. Tim Macintosh-Smith (New York: New York University Press, 2017).

4. Matthew Fontaine Maury, *Explanations and Sailing Directions to Accompany the Wind and Current Charts* (New York: Andesite Press, 2015). Originally published in 1854.

5. Lionel Casson, *The Periplus Maris Erythraei: Text with Introduction, Translation, and Commentary* (Princeton, NJ: Princeton University Press, 1989). For more on the Red Sea route in antiquity, see Nadia Durrani, *The Tihamah Coastal Plain of South-West Arabia in Its Regional Context c.6000 BC–AD 600.* BAR International Series (Oxford: Archaeopress, 2005).

6. The literature is enormous and growing rapidly. For a good summary, see Timothy Insoll, *The Archaeology of Islam in Sub-Saharan Africa* (Cambridge: Cambridge University Press, 2003), 172–177.

7. Roger Summers, *Ancient Mining in Rhodesia and Adjacent Areas* (Salisbury: National Museums of Rhodesia, 1969), 218.

8. David W. Phillipson, *African Archaeology*, 3rd ed. (Cambridge: Cambridge University Press, 2010).

9. T. N. Huffman, "Archaeological Evidence for Climatic Change During the Last 2000 Years in Southern Africa," *Quaternary International* 33 (1996): 55–60.

10. The following paragraphs rely on P. D. Tyson et al., "The Little Ice Age and Medieval Warming in South Africa," *South African Journal of Science* 96, no. 3 (2000): 121–125.

11. Peter Robertshaw, "Fragile States in Sub-Saharan Africa," in *The Evolution of Fragility: Setting the Terms*, ed. Norman Yoffee (Cambridge, UK: McDonald Institute for Archaeological Research, 2019), 135–160, offers a discussion of the issues covered in this section. See also Matthew Hannaford and David J. Nash, "Climate, History, Society over the Last Millennium in Southeast Africa," *WIREs Climate Change* 7, no. 3 (2016): 370–392.

12. Graham Connah, *African Civilizations*, 3rd ed. (Cambridge: Cambridge University Press, 2015), is a definitive summary. T. N. Huffman, "Mapungubwe and the Origins of the Zimbabwe Culture," *South African Archaeological Society Goodwin Series* 8 (2000): 14–29, is a useful starting point, updated by Robertshaw, "Fragile States in Sub-Saharan Africa," and Tyson et al., "The Little Ice Age and Medieval Warming in South Africa."

13. Peter S. Garlake, *Great Zimbabwe* (London: Thames & Hudson, 1973), is still a basic source, although somewhat outdated. Robertshaw, "Fragile States in Sub-Saharan Africa," has numerous recent references.

14. Discussion in Tyson et al., "The Little Ice Age and Medieval Warming in South Africa."

11. A Warm Snap (536 to 1216 CE)

1. Procopius of Caesarea (c. 500–c. 570 CE) was a Byzantine Greek scholar and lawyer who was intensely critical of the emperor Justinian. His *History of the Wars* is an invaluable source on the events of the early sixth century and on the Justinian Plague. Procopius, *History of the Wars* (Cambridge, MA: Loeb Classical Library, 1914), IV, xiv, 329.

2. Michael McCormick, Paul Edward Dutton, and Paul A. Mayewski, "Volcanoes and the Climate Forcing of Carolingian Europe, A.D. 750–950," *Speculum* 82 (2007): 865–895, is a fundamental source on the climate of this period and on eruptions and climate, relied on here.

3. Bartholomeus Anglicus (c. 1203–1272 CE), commonly known as Bartholomew the Englishman, was a Franciscan scholar and church official. His nineteen-book *De proprietatibus rerum* (*On the Properties of Things*) was a forerunner of the modern-day encyclopedia and was widely read. It covered a broad spectrum of subjects, including God and animals.

4. Ulf Büntgen and Nicola Di Cosmo, "Climatic and Environmental Aspects of the Mongolian Withdrawal from Hungary in 1242 CE," *Nature Scientific Reports* 6 (2016): 25606.

5. Hubert Lamb, *Climate, History and the Modern World*, 2nd ed. (Abingdon, UK: Routledge, 1995), is an excellent guide to Lamb's work. For the MCA, see his "The Early Medieval Warm Epoch and Its Sequel," *Palaeogeography, Palaeoclimatology, Paleoecology* 1 (1965): 13–37.

6. Michael Mann et al., "Global Signatures and Dynamical Origins of the Little Ice Age and the Medieval Climate Anomaly," *Science* 326 (2009): 1256–1260.

7. Ulf Büntgen and Lena Hellman, "The Little Ice Age in Scientific Perspective: Cold Spells and Caveats," *Journal of Interdisciplinary History* 44 (2013): 353–368. Sam White, "The Real Little Ice Age," *Journal of Interdisciplinary History* 44, no. 3 (winter 2014): 327–352, also offers significant insights. Büntgen and Hellman stress that the results of all this painstaking research are but provisional, for many highly technical issues await resolution. The central problem revolves around the need to calibrate meticulously collected and accurately dated proxy archives with reliable networks of instrument measurements. In general terms, however, the existing research at least provides an overall impression of climatic variations that is far more accurate than those from earlier research. The climate picture will become far more precise in future years, much of it derived from highly sophisticated, and sometimes frankly esoteric, statistical calculations that lie in specialist hands. But already we know far more about medieval climate and its vicissitudes than we did even a few years ago.

8. Ulf Büntgen et al. "Tree-ring Indicators of German Summer Drought over the Last Millennium," *Quaternary Science Reviews* 29 (2010): 1005–1016.

9. The literature on medieval agriculture is enormous. Grenville Astill and John Langdon, eds., *Medieval Farming and Technology: The Impact of Agricultural Change in Northwest Europe* (Leiden: Brill, 1997), is a valuable overview.

10. William of Malmesbury (c. 1096–1143) was a monk in southwestern England and a highly regarded historian, second only to the Venerable Bede. His *Historia Novella*, Book V, describes vineyards of the day.

11. Based on Hubert Lamb, *Climate, History and the Modern World* (London: Methuen, 1982), 169–170.

12. William Chester Jordan, *The Great Famine* (Princeton, NJ: Princeton University Press, 1996), is the definitive account of the famine, and we have relied heavily on it here. See also William Rosen, *The Third Horseman: Climate Change and the Great Famine of the 14th Century* (New York: Viking, 2014).

13. Quotes in this paragraph: Abbott of St. Vincent: Martin Bouquet et al., eds., *Recueil des historiens des Gaules et de la France*, 21:197. From Jordan, *The Great Famine*, 18.

14. Passage based on Rosen, *The Third Horseman*, 149–151.

15. Wendy R. Childs, ed. and trans., *Vita Edwardi Secundi: The Life of Edward II* (New York: Oxford University Press, 2005), 111.

16. C. A. Spinage, *Cattle Plague: A History* (New York: Springer, 2003).

12. "New Andalusia" and Beyond (1513 CE to today)

1. The Norse settlement of Greenland and subsequent voyages across to North America have been studied intensively, including superb excavations by Danish archaeologists in Greenland. See Kristen A. Seaver, *The Frozen Echo: Greenland and the Exploration of North America, ca. A.D. 1000–1500* (Stanford, CA: Stanford University Press, 1996). For L'Anse aux Meadows, see Helga Ingstad, ed., *The Norse Discovery of America* (Oslo: Norwegian University Press, 1985). Whether L'Anse is the actual place where Eirik wintered has been questioned. The controversy is unresolved.

2. Nicolás Young et al., "Glacier Maxima in Baffin Bay During the Medieval Warm Period Coeval with Norse Settlement," *Science Advances* 1, no. 11 (2015). doi: 10.1126/sciadv.1500806.

3. Brian Fagan, *Fish on Fridays: Feasting, Fasting, and the Discovery of the New World* (New York: Basic Books, 2006), offers a synthesis.

4. Sam W. White, *A Cold Welcome: The Little Ice Age and Europe's Encounter with North America* (Cambridge, MA: Harvard University Press, 2017), is the definitive source on the subject. We've relied on it heavily in writing the remainder of this chapter.

5. Climate is discussed in White, *A Cold Welcome*, 9–19. See also Karen Kupperman, "The Puzzle of the American Climate in the Early Colonial Period," *American Historical Review* 87 (1982): 1262–1289.

6. Anne Lawrence-Mathers, *Medieval Meteorology: Forecasting the Weather from Aristotle to the Almanac* (Cambridge: Cambridge University Press, 2019).

7. White, *A Cold Welcome*, 28–47, has a comprehensive discussion.

8. White, *A Cold Welcome*, 31–32.

9. Quotes in this paragraph are from White, *A Cold Welcome*, 38, 41.

10. Roanoke: Karen Kupperman, *Roanoke: The Abandoned Colony* (Lanham, MD: Rowman & Littlefield, 2007).

11. David W. Stahle et al., "The Lost Colony and Jamestown Droughts," *Science* 280, no. 5363 (1998): 564–567.

12. Richard Halkuyt, *Voyages and Discoveries: The Principal Navigations, Voyages, Traffiques and Discoveries of the English Nation*, ed. Jack Beeching. Reissue ed. (New York: Penguin, 2006). See also White, *A Cold Welcome*, 103–108.

13. Quotes by White, *A Cold Welcome*, 105.

14. This section on Jamestown is based on White, *A Cold Welcome*, chap. 6. See also Karen Kupperman, *The Jamestown Project* (Cambridge, MA: Harvard University Press, 2007), and James Horn, *A Land as God Made It: Jamestown and the Birth of America* (New York: Basic Books, 2005).

15. Stahle et al., "The Lost Colony and Jamestown Droughts," describes the tree ring research. See also T. M. Cronin et al., "The Medieval Climate Anomaly and Little Ice Age in Chesapeake Bay and the North Atlantic Ocean," *Palaeogeography, Palaeoclimatology, Paleoecology* 297 (2010): 299–310.

16. Karen Kupperman, "Apathy and Death in Early Jamestown," *Journal of American History* 66 (1979): 24–40.

17. Helen C. Rountree, *The Powhatan Indians of Virginia: Their Traditional Culture* (Norman: University of Oklahoma Press, 1989), is an invaluable source.

18. The literature is growing rapidly. For a summary, see Martin Gallivan, "The Archaeology of Native Societies in the Chesapeake: New Investigations and Interpretations," *Journal of Archaeological Research* 19 (2011): 281–325.

19. Helen C. Rountree, *Pocahontas, Powhatan, Opechancanough: Three Indian Lives Changed by Jamestown* (Charlottesville: University of Virginia Press, 2005), 64.

20. William M. Kelso, *Jamestown: The Truth Revealed* (Charlottesville: University of Virginia Press, 2018).

21. Nunalleq is known from recent excavations: Paul M. Ledger et al., "Dating and Digging Stratified Archaeology in Circumpolar North America: A View from Nunalleq, Southwestern Alaska," *Arctic* 69, no. 4 (2019): 278–390. See also Charlotta Hillerdal, Rick Knecht, and Warren Jones, "Nunalleq: Archaeology, Climate Change, and Community Engagement in a Yup'ik Village," *Arctic Anthropology* 56 (2019): 18–38.

22. Gideon Mailer and Nicola Hale, *Decolonizing the Diet: Nutrition, Immunity, and the Warning from Early America* (New York: Anthem Press, 2018), is a useful general introduction to this emerging field of research.

23. A. Park Williams et al., "Large Contribution from Anthropogenic Warming to an Emerging North American Megadrought," *Science* 368, no. 6488 (2020): 314–318. For a summary for more general readers, see David W. Stahle, "Anthropogenic Megadrought," *Science* 368, no. 6488 (2020): 238–239.

13. The Ice Returns (c. 1321 to 1800 CE)

1. Hubert Lamb and Knud Frydendahl, *Historic Storms of the North Sea, British Isles, and Northwestern Europe* (Cambridge: Cambridge University Press, 1991), is a magnificent study that describes the meteorology behind the Grote Mandreke and other tempests. Quote from p. 93.

2. Ole J. Benedictow, *The Black Death, 1346–1353: The Complete History* (Woodbridge, UK: Boydell & Brewer, 2006).

3. M. Harbeck et al., "Distinct Clones of *Yersinia pestis* Caused the Black Death," *PLOS Pathology* 9, no. 5 (2013): c1003349.

4. Boris V. Schmid et al., "Climate-Driven Introduction of the Black Death and Successive Plague Reintroductions into Europe," *Proceedings of the National Academy of Sciences* 112, no. 10 (2015): 3020–3025.

5. François Matthes, "Report of Committee on Glaciers," *Transactions of the American Geophysical Union* 20 (1939): 518–523.

6. Environmental historians have lavished attention on the Little Ice Age in recent years, so there is now rich historical material, much of it focused on the sixteenth and seventeenth centuries. Philipp Blom, *Nature's Mutiny: How the Little Ice Age of the Long Seventeenth Century Transformed the West and Shaped the Present* (New York: W. W. Norton, 2020), and Dagmar Degroot, *The Frigid Golden Age: Climate Change, the Little Ice Age, and the Dutch Republic, 1560–1720* (Cambridge: Cambridge University Press, 2018), are particularly recommended. See also Geoffrey Parker, *Global Crisis: War, Climate Change and Catastrophe in the Seventeenth Century* (New Haven, CT: Yale University Press, 2013). For ice pulses and the beginning of the Little Ice Age, see Martin M. Miles et al., "Evidence for Extreme Export of Arctic Sea Ice Leading the Abrupt Onset of the Little Ice Age," *Science Advances* 6, no. 38 (2020). doi.10.1126/sciadv.aba4320.

7. Schaller quotes cited by Blom, *Nature's Mutiny*, 30–31.

8. Described by Blom, *Nature's Mutiny*, 39–40.

9. Degroot, *The Frigid Golden Age*, is a definitive source for this section.

10. Degroot, *The Frigid Golden Age*, 130.

11. Discussion in Degroot, *The Frigid Golden Age*, 130–149.

12. Dutch engineer Cornelius Vermuyden (1595–1677) carried out drainage projects in several parts of England, including the fens in the east. His efforts were only partially successful there, until steam pumps came into use.

13. Robert Bakewell (1725–1795) was an agronomist and expert on animal husbandry, especially of sheep. He fertilized pastureland to improve grazing. His wool-bearing sheep were exported as far as Australia and New Zealand, while he was the first to breed cattle for beef, their weight more than doubling during the eighteenth century.

14. William Derham (1657–1735) was rector of Upminster near London. He had a passion for mathematics, philosophy, and science. He developed the earliest reasonably accurate measurements of the speed of sound. Quote from the "Observations upon the Spots That Have Been upon the Sun, from the Year 1703 to 1711. with a Letter of Mr. Crabtrie, in the Year 1640. upon the Same Subject. by the Reverend Mr William Derham, F. R. S," *Philosophical Transactions of the Royal Society* 27 (1711): 270.

15. For a summary of the solar minima for general readers, see Degroot, *The Frigid Golden Age*, 30–49.

16. J.-C. Thouret et al., "Reconstruction of the AD 1600 Huaynaputina Eruption Based on the Correlation of Geological Evidence with Early Spanish Chronicles," *Journal of Vulcanology and Geothermal Research* 115, nos. 3–4 (2002): 529–570.

17. Gary K. Waite, *Eradicating the Devil's Minions: Anabaptists and Witches in Reformation Europe, 1525–1600* (Toronto: University of Toronto Press, 2007).

18. This section is based on Degroot, *The Frigid Golden Age*, chaps. 2 and 3. The VOC: pp. 81–108.

14. Monstrous Eruptions (1808 to 1988 CE)

1. Francisco José de Caldas was director of the astronomical observatory of Bogotá, Colombia, from 1805 to 1810. Quote from A. Guevara-Murua et al., "Observations of a Stratospheric Aerosol Veil from a Tropical Volcanic Eruption in December 1808: Is This the 'Unknown' ~1809 Eruption?" *Climate of the Past Discussions* 10, no. 2 (2014): 1901. Whether the mysterious eruption occurred in late 1808 or 1809 is still being debated.

2. Stefan Brönnimann et al., "Last Phase of the Little Ice Age Forced by Volcanic Eruptions," *Nature Geoscience* 12 (2019): 650–656.

3. Here we have drawn on Gillen D'Arcy Wood, *Tambora: The Eruption That Changed the World* (Princeton, NJ: Princeton University Press, 2014), an excellent popular account of the eruption, as well as William Klingaman and Nicholas P. Klingaman, *The Year Without Summer: 1816 and the Volcano That Darkened the World and Changed History* (New York: St. Martin's Press, 2013).

4. Miranda Shelley, *Mary Shelley* (London: Simon & Schuster, 2018).

5. Karl Freiherr von Drais (1785–1851) was a prolific inventor who invented not only the velocipede but also the earliest typewriter with a keyboard in 1821 and even a foot-driven, human-powered railroad vehicle, the forerunner of today's railroad handcars. He renounced his aristocratic title in 1848 as a belated tribute to the French Revolution and died penniless.

6. John D. Post, *The Last Great Subsistence Crisis in the Western World* (Baltimore: John Hopkins University Press, 1977), is the definitive source.

7. Irish famine: Wood, *Tambora*, chap. 8.

8. Christopher Hamlin, *Cholera: The Biography* (New York: Oxford University Press, 2008), is a standard work.

9. Quote from Wood, *Tambora*, 97. Chapter 5 describes events in Yunnan, an account we relied on here.

10. This passage based on Wood, *Tambora*, chap. 9.

11. Thomas Jefferson, *Notes on the State of Virginia* (Chapel Hill: University of North Carolina Press, 2006). Originally published in 1784 in Paris.

12. Thomas H. Painter et al., "End of the Little Ice Age in the Alps Forced by Industrial Black Carbon," *Proceedings of the National Academy of Sciences* 110, no. 38 (2013): 15216–15221.

13. Richard H. Grove, *Ecology, Climate, and Empire: Colonialism and Global Environmental History, 1400–1940* (Cambridge, UK: White House Press, 1997).

14. *Rodney and Otamatea Times, Waitemata and Kaipara Gazette*, August 14, 1912.

15. Peter Brimblecombe, *The Big Smoke: A History of Air Pollution in London Since Medieval Times* (Abingdon, UK: Routledge, 1987). See also Stephen Halliday, *The Great Stink of London: Sir Joseph Bazalgette and the Cleansing of the Victorian Metropolis* (Stroud, UK: Sutton, 2001).

16. C. S. Zerefos et al., "Atmospheric Effects of Volcanic Eruptions as Seen by Famous Artists and Depicted in Their Paintings," *Atmospheric Chemistry and Physics* 7, no. 15 (2007): 4027–4042; Hans Neuberger, "Climate in Art," *Weather* 25, no. 2 (1970): 46–56.

17. James Hanson, congressional testimony, June 23, 1988.

15. Back to the Future (Today and Tomorrow)

1. Raphael Meukom et al., "No Evidence for Globally Coherent Warm and Cold Periods over the Preindustrial Common Era," *Nature* 571 (2019): 550–554.

2. "The past is a foreign country": L. P. Hartley, *The Go Between* (New York: New York Review Book Classics, 2011). David Lowenthal, *The Past Is a Foreign Country*, 2nd ed. (Cambridge: Cambridge University Press, 2015), is a recent discussion.

3. The most effective way to understand the dimensions of global warming and potential solutions lies in Paul Hawken, ed., *Drawdown: The Most Comprehensive Plan Ever Proposed to Reverse Global Warming* (New York: Penguin Books, 2017). The essays in this remarkable book offer ideas and potential solutions that are sometimes breathtakingly simple but always forward-looking.

INDEX

LESLEY NEWHART

Brian Fagan is one of the world's leading archeological writers and an internationally recognized authority on world prehistory. He is a Distinguished Emeritus Professor of Anthropology at the University of California, Santa Barbara, and the author of several widely read books on ancient climate change. He has lectured about the subject to audiences large and small throughout the world. Among his latest books is *Fishing: How the Sea Fed Civilization* (2018).

JACOB HILLIER

Nadia Durrani is a Cambridge University–trained archeologist and writer, with a PhD from University College, London, in Arabian archeology. She is founder and editor of *THE PAST* (www.thepastmagazine.com) and former editor of both *Current Archaeology* and *Current World Archaeology* magazines. She is coauthor of a portfolio of textbooks with Brian Fagan and also the trade books *What We Did in Bed: A Horizontal History* (2019) and *Bigger Than History: Why Archaeology Matters* (2019).

PublicAffairs is a publishing house founded in 1997. It is a tribute to the standards, values, and flair of three persons who have served as mentors to countless reporters, writers, editors, and book people of all kinds, including me.

I. F. STONE, proprietor of *I. F. Stone's Weekly*, combined a commitment to the First Amendment with entrepreneurial zeal and reporting skill and became one of the great independent journalists in American history. At the age of eighty, Izzy published *The Trial of Socrates*, which was a national bestseller. He wrote the book after he taught himself ancient Greek.

BENJAMIN C. BRADLEE was for nearly thirty years the charismatic editorial leader of *The Washington Post*. It was Ben who gave the *Post* the range and courage to pursue such historic issues as Watergate. He supported his reporters with a tenacity that made them fearless and it is no accident that so many became authors of influential, best-selling books.

ROBERT L. BERNSTEIN, the chief executive of Random House for more than a quarter century, guided one of the nation's premier publishing houses. Bob was personally responsible for many books of political dissent and argument that challenged tyranny around the globe. He is also the founder and longtime chair of Human Rights Watch, one of the most respected human rights organizations in the world.

• • •

For fifty years, the banner of Public Affairs Press was carried by its owner Morris B. Schnapper, who published Gandhi, Nasser, Toynbee, Truman, and about 1,500 other authors. In 1983, Schnapper was described by *The Washington Post* as "a redoubtable gadfly." His legacy will endure in the books to come.

Peter Osnos, *Founder*